Born and Made

IN·FORMATION *Series*

Series Editor
Paul Rabinow

A list of titles in the series appears at the back of the book

Born and Made

An Ethnography of Preimplantation Genetic Diagnosis

Sarah Franklin
and
Celia Roberts

PRINCETON UNIVERSITY PRESS

PRINCETON AND OXFORD

Library of Congress Cataloging-in-Publication Data
Franklin, Sarah, 1960–
Born and made : an ethnography of preimplantation genetic diagnosis /
Sarah Franklin and Celia Roberts.
p. ; cm. — (In-formation series)
Includes bibliographical references and index.
ISBN-13: 978-0-691-12192-5 (hardcover : alk. paper)
ISBN-10: 0-691-12192-3 (hardcover : alk. paper)
ISBN-13: 978-0-691-12193-2 (softcover : alk. paper)
ISBN-10: 0-691-12193-1 (softcover : alk. paper)
1. Preimplantation genetic diagnosis—Social aspects—Great Britain.   2. Preimplantation
genetic diagnosis—Moral and ethical aspects.   I. Roberts, Celia, 1968–   II. Title.
III. Title: Ethnography of preimplantation genetic diagnosis.   IV. Series.
[DNLM: 1. Preimplantation Diagnosis–ethics–Great Britain.   2. Genetic Diseases,
Inborn–diagnosis–Great Britain. QZ 50 F834b 2006]
RG628.3.P74F73 2006
618.2'075—dc22   2006021714

British Library Cataloging-in-Publication Data is available

In memory of Margaret Stacey & Dorothy Nelkin

for their inspirational legacies and their distinguished scholarly contributions to the social study of biomedicine

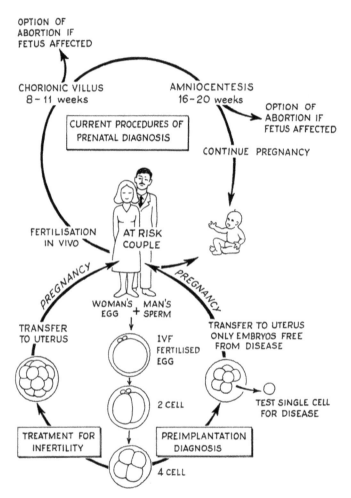

OPTION OF
ABORTION IF
FETUS AFFECTED

CHORIONIC VILLUS
8 – 11 weeks

AMNIOCENTESIS
16 – 20 weeks

CURRENT PROCEDURES OF
PRENATAL DIAGNOSIS

OPTION OF
ABORTION IF
FETUS AFFECTED

CONTINUE PREGNANCY

FERTILISATION
IN VIVO

AT RISK
COUPLE

PREGNANCY

PREGNANCY

WOMAN'S + MAN'S
EGG   SPERM

TRANSFER
TO UTERUS

TRANSFER TO UTERUS
ONLY EMBRYOS FREE
FROM DISEASE

IVF
FERTILISED
EGG

2 CELL

TEST SINGLE CELL
FOR DISEASE

TREATMENT FOR
INFERTILITY

PREIMPLANTATION
DIAGNOSIS

4 CELL

"Options for couples at risk." An early diagram of PGD from the archives of
Marilyn Monk.

Contents

Figures

Acknowledgments

The research for *Born and Made* was supported by two of the national research councils in the United Kingdom, the Economic and Social Research Council and the Medical Research Council, under their cosponsored Innovative Health Technologies (IHT) Programme (research award reference number L218252036), whose director, Andrew Webster, offered his support and encouragement throughout this project. Sarah Franklin was the principal investigator, Peter Braude and Tony Rutherford were coapplicants, and Celia Roberts was the research associate for this social-scientific study of preimplantation genetic diagnosis (PGD), which was funded from January 2001 to August 2002 and continued until 2005. Peter Braude and Tony Rutherford provided assistance in designing and securing funding for the project, as well as facilitating all the formal project requirements in London and Leeds, including the lengthy process of securing ethical approval. They were also instrumental in building stronger links between the project team and the wider scientific community. Michal Nahman at Lancaster University and David Reubi at the London School of Economics also provided project assistance in the form of administration and assistance securing permissions for the illustrations reproduced in *Born and Made*. Both authors conducted interviews and site visits. Celia Roberts organized the project archive, oversaw the transcription of tapes and organized the project data by themes. These themes are the basis for chapters 3, 4, and 5, which are coauthored. Sarah Franklin authored the preface, introduction, conclusion, and chapters 1, 2, and 6, which cover the history of PGD, its "accountability," and its future. She prepared the final manuscript of *Born and Made* during 2003–4 at the University of Sydney with the benefit of a research sabbatical generously provided by Lancaster University, and completed the revisions in the Department of Sociology at the London School of Economics during 2004–5. Both authors would like to thank the three readers for Princeton University Press who provided helpful critical feedback, suggestions, and advice on the original manuscript.

This study could not have been funded, undertaken, or completed without the generous sponsorship of many very busy people, and both authors would particularly like to thank Jenny Caller and Karen Miller, the PGD coordinators in London and Leeds, who provided the core logistical assistance for all aspects of the fieldwork for this study. Special thanks are also due to the Guy's and St. Thomas' ACU scientific director, Soo Pickering, whose talents in the art of scientific explanation are beyond description. To all the members of the London and Leeds PGD teams many thanks are due for patience, good humor, and very helpful feedback. In particular, thanks to Alan Handyside, Helen Picton, Jan Hogg, and Fiona Robson at Leeds for taking time out to explain specific and technical aspects of PGD in practice, and to Alison Lashwood, Cindy Zaitsoff, Cheryl Black, and Jan Grace for their contributions to this research in London. Alan Handyside, Anne McLaren, Marilyn Monk, Ruth Deech, Virginia Bolton, Kay Elder, Peter Braude, and Juliet Tizzard also provided very helpful background on PGD's scientific development and "parliamentary career." Ginny Squires, Debbie Jaggers, and Chris O'Toole of the Human Fertilisation and Embryology Authority helpfully explained the workings of the Authority and provided details about the history of regulation, legislation, and policy in the United Kingdom during a busy period in their working lives. Lastly, thanks are due to Onora O'Neill, whose Reith lectures about accountability were especially helpful for interpreting project data about uncertainty.

Above all, this study has benefited from the individuals and couples who agreed to be interviewed about their experiences of undergoing PGD, and who allowed their consultations and treatment to be observed in clinic. Every effort has been made to anonymize the PGD patients who appear in *Born and Made* out of respect for their privacy and that of their families, friends, children, and colleagues. Without their exceptionally generous hospitality and willingness to speak so openly and courageously about very personal and often disturbing experiences, in the midst of difficult and time-consuming treatment, this research could not have taken place. The extent to which this book can enable greater recognition of patients' experiences of PGD, the dilemmas they face, and the reasons why these should concern us all is largely the result of the sometimes shocking honesty with which both PGD patients and professionals in this field expressed themselves during this study.

This project benefited from considerable assistance and support from academic colleagues during its preparation, undertaking, and the writing of this monograph. In addition to the social-scientific and medical audiences who responded to portions of this book as work in progress, special thanks are due to everyone who read and commented on sections of *Born and Made*, including Sara Ahmed, Virginia Bolton, Peter Braude, Jenny Caller, Robert Edwards, Kay Elder, Monica Konrad, Alison Lashwood, Marilyn Monk, Anne McLaren, Bronwyn Parry, Nikolas Rose, Bob Simpson, Pat Spallone, Karen Throsby, Clare Williams, and Kristin Zeiler. Marilyn Strathern characteristically provided crucial advice and much-needed guidance at some particularly "tight junctions."

Sarah Franklin would like to thank the Department of Gender Studies at the University of Sydney, and in particular Elspeth Probyn, for much-appreciated hospitality during the writing of *Born and Made* in 2003–4, and the Department of Sociology at the London School of Economics (LSE) for logistical and financial support for completing the final manuscript during 2004–5. Both authors would like to thank David Reubi from the BIOS Centre at the LSE for his excellent work securing permissions for the illustrations used in *Born and Made*, as well as several members of the Department of Sociology at Lancaster University, especially Claire O'Donnell, Pennie Drinkall, Cath Gorton, Karen Gammon, Joanne Bowker, Dan Shapiro, and Lucy Suchman, as well as other members of the Lancaster community, in particular Maureen McNeil, Maggie Mort, and Imogen Tyler. Special acknowledgments are due to Hilary Graham for her assistance preparing the successful grant application for this study, to Jackie Stacey for insisting that nothing can ever be "just" descriptive, and to Sara Ahmed for detailed comments and insightful questions. Celia Roberts would additionally like to thank Adrian Mackenzie, John Law, Cathy Clay, Hilary Roberts, Hugh Roberts, Tom Roberts, Suzanne Fraser, kylie valentine, Barbara Caine, and Cathy Waldby.

A Note on Identification of Speakers in the Text

As with any study of this kind, and especially when very intimate aspects of personal life are concerned, it is not always possible to know what is the ideal balance between acknowledging individual speakers in the text and concealing their identity. It is also not possible to know how much information to reveal about the identity of speakers who are named. The standard conventions of anonymizing as fully as possible everyone who is quoted about his or her personal experience of undergoing PGD are used in this book, despite the fact that this has some inevitable drawbacks.

For example, interviewees are not introduced as "Mary, a thirty-two-year-old schoolteacher from Newcastle," or "Martin, a Cambridge-educated scientist," as such details are potentially identifying, much as the lack of them can seem somewhat impersonal. The very small number of people undergoing PGD requires that anonymization must be interpreted broadly in this context. Participants who spoke about their personal experiences of PGD are referred to as "patients," as "couples," or by their pseudonyms, despite the fact that they are, in a sense, not themselves "patients," and despite the way in which the use of "couple" can obscure important distinctions between the different partners' roles, opinions, and so forth. These are the terms that patients and couples use to describe themselves, and although they have shortcomings, they are the terms employed in this book. The names of places, hospitals, medical staff, friends, or family members were removed from all the transcripts from which quotations were taken, and the names of couples' children were changed, whether they were living or deceased. These were agreed-upon protocols at the outset of the study and have been maintained, and adapted, as consistently as possible.

In field notes and interview transcripts that did not concern personal experiences of undergoing PGD, including discussions with PGD staff in clinics, speakers are not anonymized, and, by agreement with the ethics committees at both hospitals, these speakers are identified by name and are semiformally identified in terms of their professional role and expertise in a footnote, usually when they are first introduced within the text. The aim of this practice was to assist readers in identifying "who's who" in the text, and thus to contribute to a sense of the "world of PGD." To ensure that nothing was misquoted, all descriptions and quotations were shown to the relevant people to be checked and, in some instances, were changed to reflect the actor/speaker's sense of his or her original meaning.

While this strategy has the disadvantage of creating two sets of voices—those that belong to named (expert, professional) participants and those that are concealed by (patient) anonymity—this was the most preferable alternative for the reasons described above, in respect to the privacy not only of patients but also of their children, families, friends, and colleagues. Moreover, this strategy was mutually agreed upon in advance, and was the basis for the ethical approval for this study (see further comment on the advantages and disadvantages of quoting named and anonymized participants in chapter 2, "Studying PGD").

All proceeds from sales of this book have been donated to support the Great Ormond Street Hospital in London and other medical charities for children.

Preface

This book tells the story of a specific technique in a particular country and during a distinct historical period, but it addresses a general theme—of how to account for the social dimensions of new biomedical technologies. The technique of preimplantation genetic diagnosis (PGD) is of interest because it is the first to bridge between the effort to "assist" human reproduction and the ability to intervene in human heredity, thus extending the helping hands of medical science into the innermost workings of early human life. It has also played a prominent role in public debate about reproductive futures and regulation, where it is often referred to as the "designer baby technique." Built on the platform of in vitro fertilization (IVF), PGD extends the reach of both medicine and science, conjoining them in the effort to increase reproductive choice, by attempting to alleviate the suffering caused by genetic and chromosomal pathology at a stage where the embryo is still "in glass." Thus, PGD belongs to a long history of human and animal reproductive and hereditary management, while—in the age of the human genome project, increasingly powerful molecular diagnostics, and sophisticated techniques of microsurgery on embryos—it presents new kinds of genetic choice, and thus new questions about the governance of "life itself."

The United Kingdom has played a significant role in the postwar drama surrounding reproductive and genetic intervention in large part because of its leading scientific role in these fields and its vibrant climate of experimental embryology—often geared to questions of agriculture and public health—in the postwar era, coincident with the establishment of its National Health Service (NHS), the largest of its kind in Europe.

Embryo transfer, embryo biopsy, embryo freezing, IVF, PGD, and cloning of a higher vertebrate from an adult cell are but a selection of Britain's long list of scientific "firsts" in the fields of reproductive biomedicine, embryology, and developmental biology. Another British "first" was the Human Fertilisation and Embryology Act of 1990, establishing a comprehensive system of regulation for both science and medicine where they came into contact with the human embryo. The introduction and chapter 1 of this book explore this connection—between the promotion of a climate of scientific innovation in the United Kingdom and the comprehensive and sustained effort to produce legislation—in greater depth. What is significant is the role that PGD played in this process and, as later chapters explore, the role of PGD in the ongoing transformation of the relationship between science, medicine, technology, legislation, and public debate in the United Kingdom. Why PGD? Why the United Kingdom? Why does this particular story matter?

Three primary questions receive thorough examination in this book, which is an attempt to contribute to the debate about reproductive and genetic "design" from the vantage point of social science. First is how the question of being both born and made is represented, inhabited, communicated, celebrated, and decried. The second question is how to account for this mélange of "deeply felt" opinion on matters of life and death, and how to make sense of its diversity. The third is to ask, on the basis of "the social life of PGD" in the United Kingdom, how this technique, and its ilk, can be made accountable—to its users and providers; to those it fails as well as those who succeed; for those who oppose it as well as those who celebrate its accomplishments and its potential in the future; and for those who have no direct connection to it as well as those who do.

The title of this book refers to a simple dynamic that is at once obvious and perplexing, namely that we are both "born" and "made." At one level this refers to the changes that have affected "the facts of life" insofar as technological "assistance" to the process of conception has made it more explicitly contingent: reproduction can now be "achieved," and particular types of reproduction can be deliberately chosen to a degree never witnessed before (Franklin 1995b, 1997). The paradox made more explicit through this change is visible in how it pits the view that biological identities are "given" to us by our genetic origins against the opposite view—that we can, and sometimes have an obligation to, remake who we are by intervening in the basic biology of the beginning of life.

This book explores this dynamic, which one astute commentator suggested might better be described as being "made and born," in light of one of its important predecessors, namely being born and bred. In this idiom can be found an older example of the duality of identities, such as those of lineage or pedigree, which are both "given" or "fixed" and amenable to change, adaptation, and remaking or "assistance." As Jeanette Edwards argues in her widely cited monograph, *Born and Bred: Idioms of Kinship and New Reproductive Technologies in England* (2000), "the power of Born and Bred kinship [is that] one or other— born or bred—can be brought into focus, momentarily screening out, but never fully replacing, the other" (246). This "digital" quality of kinship and, increasingly, "biological" identity *to switch back and forth* is necessary because the meanings of birth, origin, parenting, or the "making" of new offspring change across different contexts, as Charis Thompson also argues in her account *Making Parents* (2005). This book argues that an important new duality of the contemporary era, and not only in Britain, is the duality of being born and made. The argument here thus follows that of Marilyn Strathern's classic account of English kinship, in which she argues that the identities we acquire through "biological relations" are always being remade, because they are never singular but "conglomerate," plural, indeterminate, and asymmetrical (1992b, 83–84; 2005).

This book is itself both born and made as a result of the fact that the role of the social sciences in critical debate about the implications of new reproductive and genetic technologies has both expanded and gained momentum. The result of an increasing amount of social-scientific research in the areas of human reproduction and genetics, as well as stem cells and cloning, is that new kinds of "interliteracy" have emerged between practitioners in diverse disciplines, such as developmental biology, clinical genetics, embryology, anthropology, sociology, and bioethics. These interliteracies also connect scientists and clinicians to policy makers, representatives of government to social scientists, and patient groups to journalists in a host of emergent alliances that together comprise new forms of "biological citizenship" (Ginsburg and Rapp 1995; Novas and Rose 2000; Petryana 2002; Rose 2001, 2006).

In some important ways that this book tries to explore through examining the case of PGD, these emergent cross-disciplinary conversations reflect changes in the relationship between science and society more generally. A hackneyed phrase, "science and society" remains a vague and often superficially formulaic subgenre of public debate. Public events featuring panels of scientific experts, philosophers, theologians, medical practitioners, patient groups, policy specialists, and the like have become a familiar format of contemporary engagement with issues such as "designer babies," "reproductive cloning," or "the promise of stem cell research." Such events often involve a kind of roll call of professional expertise, frequently accompanied by poignant testimony from those whose suffering might be, or has been, alleviated by therapeutic advances and

discoveries. The promise of an improved future, and the threat of the reverse, are the antipodean poles that frequently frame such occasions, like
opposing question marks poised for a tug-of-war to win public sympathy, understanding, and support.

However, as Sheila Jasanoff argues in her comparative analysis of the
widely divergent styles of governance of biomedicine and biotechnology
in the United States, the United Kingdom, Germany, and the European
Union as a whole, many debates about "science and society" continue to
oversimplify and overrationalize the issues at stake, often using implicit
models of both "science" and "society" that rely more on stereotypes
than on rigorous analysis. In these debates, both "science" and "society"
become reductive terms that obscure precisely what needs more clarification. For Jasanoff, the overreliance on specific events, such as the birth
of Dolly the sheep, as starting points for discussions of the role of science
in society obscures the contingencies built into how such events are
interpreted—which she describes in terms of "framing." As she points
out, frames are both contingent and determined: they are the vehicle
both for predetermined patterns of habitual response (e.g., Dolly is evidence of science "racing ahead") and for divergent responses (not everyone will frame the issue of cloning in the same way). As she notes: "In
contrast to the notion of political agenda-setting, which takes for
granted the shape of political issues, framing implicitly makes room for
the contingency of social purposes and the partiality of the imaginative
space that is carved out for political action in any society. Not every culture that looks at a set of events necessarily frames them in the same
way. Things within the frame for one set of actors . . . may fall out of the
frame for others" (2005, 25).

As Jasanoff also points out, the advantage of a model such as "framing" over one that relies on more conventional definitions of objectivity
and rationality is that it can account for the partiality and plurality of
how people define the salient facts and values, or rights and wrongs, in
complex debates over contemporary biopolitical topics such as euthana-
sia, regenerative medicine, or embryo research. Frames may partly overlap or diverge, they may stay the same or change, and they may be more
or less recognizable or comprehensible to other social actors. In this way,
framing "provides an effective way of accommodating the solidity as
well as the interpretive flexibility of the worlds in which policy gets
made," which, according to Jasanoff, may be a particularly important
means of gaining purchase on the complex worlds of social meaning that
both shape and are shaped by technological change.

However, if "framing" is an important means of accounting for the
plurality of social patterns that shape public responses to issues such as

cloning, stem cells, or "designer babies," we need to know how frames work: "we need to know about the diversity of material with which they are constructed, how they achieve taken-for-granted status, and what happens to make frames change," argues Jasanoff (2005, 25). To do this, we need both an accumulation of specific studies and better means of comparing them, which are tasks to which this study contributes. Using ethnographic methods, whereby researchers immerse themselves in a range of different contexts to collect data about a particular object of inquiry, "following it around" to build up a kind of hyperstack of definitions, images, representations, testimonies, descriptions, and conversations, the aim of this book is to document the "social life of PGD" during a key point in its development by examining how it is framed from a variety of different sites and perspectives. In addition to working closely with two PGD teams, in London and in Leeds, attending scientific conferences and public consultation events, sitting in on parliamentary debate, and learning the history of PGD, forty-five interviews were conducted for this study with scientists, clinicians, policy makers, parliamentarians, the media, and PGD patients. Chapter 2, on method, chronicles the research involved in this data collection exercise, and the rest of the book attempts to distill some of this material into a more concentrated product in the form of analysis. The result is the opportunity to learn about PGD both from those who praise or condemn it from afar and from those who inhabit the "topsy-turvy world of PGD."

As the "designer baby" technology, PGD is often associated with genetic enhancement and the "brave new world" of made-to-order offspring. Indeed, the idiom "designer baby" is itself one of the most powerful framing devices through which PGD is interpreted by those who view it from "far away," and one of the most resented interpretations of PGD by those "up close." This creates a paradox this book explores in some depth. For while the "designer baby" is in many respects the core symbol of the fear that increased genetic control coupled with powerful parental desires in a highly stratified consumer economy will pervert the course of human reproductive futures, such a view contrasts with the relatively slow development of PGD, amidst almost constant deliberation and debate, over the past two decades. It is, indeed, ironically because PGD has affected *so few people* directly that its association with "designer babies" remains all but taken for granted.

PGD began, as most reproductive technologies do, as an experimental method in the field of developmental biology, using animal models. Although it achieved its first clinical successes in the United Kingdom in 1990, few clinics were able to offer PGD, owing to its enormous medical, scientific, and technical complexity, requiring substantial interdisciplinary expertise and a high level of disciplined coordination. By the time this study began in 2001, more than a decade following the first successful clinical application of PGD in London in 1990, only four clinics were licensed in the United Kingdom to practice this technique. Unlike IVF, which was also developed in the UK but which expanded very quickly from the late 1970s onward, resulting in more than two million IVF children being born worldwide by 2005, PGD has developed at a much slower pace.

So why are PGD and the "designer baby" the symbol of too much genetic control released into the hands of choosy would-be parents who desire "perfect" blond-haired, blue-eyed children when, as the following chapters show, this is nothing like what PGD patients describe as their desires, nor what PGD professionals define as their aims? Why is PGD linked to the idea of "design"—a pejorative term that, despite its inaccuracy, increasingly dominates debates over reprogenetic technologies? Isn't IVF a much more significant "runaway" technology? After all, two million children have been born from IVF in a single generation, and in some countries, such as Denmark, children born from IVF make up as much as 5 percent of the birthrate. Despite the lack of significant follow-up studies on children born of IVF, and some of its known liabilities, such as high rates of multiple births, IVF somehow escapes the "designer" stigma associated with PGD.

As the following chapters demonstrate, PGD's "designer" connotations are highly contested among those who undergo the technique, or who work in PGD centers—almost all of whom vigorously reject the "designer baby" moniker, which is seen as a misleading and harmful stereotype. So why has it stuck? Is it because IVF is more easily "naturalized" and thus "normalized" as an imitation of "the real thing," merely "helping" nature do *what it would have done anyway*? Whereas the whole point of PGD is precisely *not* to give nature a helping hand but, rather, to prevent it doing what it might have done "by itself"? Is the intervention involved in PGD controversial because it is genetic, whereas in IVF the main focus of "redesign" is female hormone levels?

PGD requires successful IVF with a high embryo yield, sophisticated embryo biopsy, and supersensitive, high-speed diagnosis of the genetic contents of a single cell nucleus. PGD can thus be seen to be extending, or building on, the platform of IVF by adding a genetic dimension—and this is in part why "designer babies" appear to be more controversial than those born through "normal IVF"—whose births are now, on the whole, almost routine. However, the real difference between IVF and PGD is their clientele: PGD is pursued for a different purpose—to deliberately prevent initiating a pregnancy affected by a serious or lethal genetic disorder.

While it is important to distinguish between IVF and PGD, they are both increasingly intertwined with other branches of biomedicine and bioscience, and in particular with the derivation of human embryonic stem cells, genomics, and cellular reengineering. Amidst almost daily accounts of new developments in these busy fields of bioscientific and biomedical innovation, it is increasingly clear that IVF and PGD are part and parcel of a worldwide effort to unravel the molecular and biochemical architecture of embryonic development, cellular differentiation, and the "pluripotency" of the inner cell mass. Together, these developments are what are signified by this book's title, *Born and Made*. For although PGD may be the technique that is currently associated with the idea of redesigning human origins, this process has many antecedents and is now, irrevocably, part of how human reproduction is understood.

If, however, it might be said that the fact of being born *and made* has been naturalized and normalized to a degree by IVF, the extent to which human reproduction and heredity should be "redesigned" is one of the most urgent political questions of the early twenty-first century—and one on which there is an enormous amount of debate. For example, it is a striking feature of both the history of IVF and the more recent emergence of other branches of reproductive biomedicine, such as stem cell derivation and therapeutic cloning, that they are very differently debated, practiced, and governed in different parts of the world. In contrast to Britain, Sweden, Denmark, and Israel—where IVF and PGD receive significant public support and investment, alongside stem cell research and the "cloning" of human embryos—Germany, Italy, Austria, and the United States have more restrictive policies. Germany and Italy place strict restrictions even on IVF, whereas the United States has prominently rejected stem cell research funding at the federal level and lobbied for a ban on human cloning within the United Nations.

The question of what can be learned by tracking PGD as a connecting thread must be told from the perspective of a particular fabric, but that is not to say that the connections are exclusively local. Indeed, it is the particularities that matter most when we are attempting to characterize something for the first time, when we are beginning to learn its contours and trying to plumb its depths. As an embryologist working at a micromanipulator must manually raise and lower the focal plane to compensate for its lack of depth, in order to explore the three-dimensional contours of the embryo, and by so doing learns to see these contours when they are always, by definition, impossible to capture visually from any singular location, so too the layering of social perception, always in one place but always derived from many, has to be cumulative to become more fully comprehensible. The desired insight here is of this kind—specific and particular, yet part of a larger project of accumulating sufficient detail to begin to develop a new species of characterization—one that might be more fully accountable to the social complexity of being born and made.

Born and Made

Introduction
Babies by Design?

Couples (or lone mothers) will soon be able to sort through a collection of embryos and select for a place in our midst those with the most desirable gene profiles. And why indeed would those parents in our age of excess, who can afford to press on their offspring all the advantages that health and education can afford, choose to deny them any available genetic privileges at birth?
—Walter Gratzer, "Afterword" to James D. Watson, *A Passion for DNA: Genes, Genomes and Society.*

I can understand it is a very grey area. Because obviously . . . we've used PGD [preimplantation genetic diagnosis] because we didn't want to have another child that was going to die within 12 months. But I mean, . . . at what point do you draw the line? At a child that dies at 2 years, 5 years, 10 years, 20 years, 30 years? Where? . . . What conditions are we going to allow PGD to be used for? . . . I don't know where the line should be drawn.
—Anne, PGD patient

One of the late twentieth century's most infamous offspring, the "designer baby" has become, alongside the clone, a familiar figure in debates about new reproductive and genetic technologies in what has come to be known as the "postgenomic" era. Like the iconic image of the "test-tube baby" that preceded it, the "designer baby" signifies a disturbing mixture of newfound biogenetic control, consumer demand, and parental desire. An ambivalent figure, the designer baby is at once celebrated as a medical-scientific breakthrough and decried as an example of "science gone too far." Alongside media celebrations of joyful parents enabled to have a healthy child with the assistance of modern medical technology are ominous depictions of too much choice and control over reproduction. Above all, the "designer baby" symbolizes and embodies the question of limits: How far should science be allowed to go? Who should decide, and how? How will "society" be protected against the possibility of reproductive biomedicine "going too far"? And how can the needs of "desperate" individual parents and families be balanced against the needs of society "as a whole"?

The very form of these familiar questions—full of vague, uncertain agencies and urgencies—is indicative of the difficulty of finding adequate language to address what are increasingly prominent and, in the twenty-first century, increasingly intimate questions about reproductive and genetic choice and control. This same question of finding adequate language is evident in the very term "designer baby," which, as the contrast between the two opening epigraphs demonstrates, evokes very different images, from choosy parents seeking "genetic privileges" for their children to cautious parents who feel obligated to avoid future harm to their potential offspring. For the last half century new forms of reproductive and genetic intervention—from the early research on the manipulation of mammalian embryos in the 1950s to the development of in vitro fertilization (IVF) in the 1970s, and its rapid expansion since—have been the subject of increasing social, ethical, and political concern, almost always framed as a conflict between the need for greater medical-scientific progress and the risks of "going too far." But who are the "we" of the "society" in need of protection against "science" going too far? What is "too far," and who speaks for "science"? How is the future being imaged and imagined in such debates, and how will limits be devised and implemented? Whose interests will prevail, and who is put at greater risk by medical-scientific intervention into not only reproduction but, increasingly, heredity?

Born in Britain

It is a measure of the degree of controversy surrounding IVF, embryo research, and more recently cloning and human embryonic stem cell derivation that, from their inception, such techniques have been topics that generate widely divergent responses, not only at the level of individual opinion, but equally at the level of national governance and policy (Banchoff 2004; Jasanoff 2005). Britain has in many respects been at both the center and the forefront of the controversies surrounding a cluster of new technologies associated with reproductive biomedicine, not only because so many "firsts" were born in Britain but also because it has played a more substantial role than any other country in the creation of rigorous legislation and policy strictly limiting technological manipulation of "human fertilisation and embryology" (Gunning 2000; Jackson 2001; Morgan and Lee 1991).

Home to the world's first test-tube baby (1978), first clinical use of PGD (1990), and first cloned "higher" vertebrate (1996), as well as being one of the leading countries involved in human embryonic stem cell derivation and "banking," Britain has simultaneously pursued an arduous

2

course of legislative and regulatory innovation to establish a uniquely robust-but-flexible system of laws and codes of conduct. These are backed up by criminal law, enforced through a licensing body, and subject to constant revision, while being bound by the "will of Parliament" expressed in the 1990 Human Fertilisation and Embryology Act.

This much-admired and widely emulated system of governance has its roots in the Committee of Inquiry chaired by the Oxbridge philosopher Mary Warnock in the 1980s, which both provided the rationale for and then successfully established the very liberal but highly regulated climate of reproductive biomedicine that is increasingly seen to be distinctively British (Warnock 1985). Significantly, however, and as Mary Warnock herself was at pains to make clear in her report, the basis for the committee's recommendations, while informed by perspectives from moral philosophy, theology, and bioethics (fields in which several of the committee members held prestigious positions), *was essentially sociological rather than philosophical*. Conspicuously, and often controversially, eschewing the perennial jousting match over "the moral status of the human embryo," the primary question guiding Warnock was "what kind of society can we praise or admire? In what sort of society can we live with our conscience clear?" (1985, 3). The first premise of Warnock's approach was "to take very seriously the . . . wide diversity in moral feelings" and to determine which feelings were most strongly held in common. For Warnock, the most strongly shared social consensus was that "people generally want *some principles or other* to govern the development and use of the new techniques. There must be some barriers that are not to be crossed, *some fixed limits*, beyond which people must not be allowed to go. . . . The very existence of morality depends on it. A society which had no inhibiting limits . . . would be a society without scruples, and this nobody wants" (1985, 2, emphasis added).

"The Warnock position," as sociologist Michael Mulkay has argued, was the outcome of an effort to acknowledge fundamentally opposing views and find the path of greatest social consensus among them (Mulkay 1997). The report's primary recommendation—to establish a licensing authority that would provide overall regulation—was based on an original "social contract" devised by Warnock, through which maximum scientific innovation would be encouraged, so long as it was subject to the very strictest levels of government regulation. This "Warnock strategy" in effect substituted robust regulatory infrastructure for "principalism," thereby establishing a pattern that has prevailed ever since in Britain.[1]

---

[1] According to philosopher John Harris's critique of the Warnock Report, "the crucial questions are fudged, or rather are never addressed" (1985, 130), and the committee

FIGURE I.1. This portrait of the philosopher Mary Warnock from the National Portrait Gallery in London was taken by the photographer Steve Pyke in 1990, the year in which the UK's famous Human Fertilisation and Embryology Bill was enacted, and the first year in which PGD was clinically practiced. Photograph © Steve Pyke.

It is a strategy that has been widely acclaimed, and more often emulated internationally, than any other.

"treat[ed] people's expression of strong feeling as moral, whatever sort of world they are likely to produce . . . simply because they are feelings expressed sincerely about important moral matters" (1985, 132). He adds that "it is one thing to conclude that morality depends on barriers, and quite another to assume that barriers make morality" (1985, 132).

Importantly, it is a strategy that is based not on absolute values of right and wrong but on the "bottom line" of deliberation within an established legislative system. As Warnock herself described the process of her committee in retrospect, writing in her 2003 publication *Nature and Morality*:

> An absolutely central consideration in the work of [our] committee . . . was the difference between what one might personally think was sensible, or even morally right, and *what was most likely to be acceptable as a matter of public policy*. . . . Time and again we found ourselves distinguishing not between what would be right or wrong, but between what would be acceptable or unacceptable. (Warnock 2003, 98–99, emphasis added)

Instead of a resolution of moral differences based on philosophical principles, which Warnock has described as "impossible" (2003, 99), she chose "to try to assemble a coherent policy which might seem, if not right, then at least all right, to the largest possible number of people" (2003, 99).[2]

It is primarily for this reason—that the basic principle informing British regulation of new reproductive and genetic technologies is *to achieve workable and sustainable policy*—that the British government has, from 2000 onward, increasingly funded social-scientific research into the area we might call "reprogenomic studies" or the social study of biomedicine. Bioethics too is moving in this direction, as a new genre of empirically based "context-specific ethics" increasingly replaces debates over utilitarianism versus consequentialism, or dignitarianism versus liberalism (Haimes 2000; Hedgecoe 2003). The research presented in this book was funded under just such an initiative, coordinated by the Economic and Social Research Council (ESRC), which in 2001 commissioned a comprehensive program of more than forty studies on the topic of "innovative health technologies" (Brown and Webster 2004). The two-year study on which this book is based also benefited from funding by the ESRC Genomics Programme, inaugurated in 2002, under which more than thirty million pounds (US$60 million) was committed to social-scientific study of genomics, stem cells, and reproductive biomedicine. This vast investment in the social science of biomedicine in Britain dwarfs that of any other country and is unprecedented. It is a further extension of the "Warnock strategy," through which an essentially

---

[2] In her 2003 publication *Nature and Morality: Recollections of a Philosopher in Public Life*, Mary Warnock describes the crucial importance of her realization that "the language of 'right' and 'wrong' was inflammatory": it was antisocial, and "it sounded arrogant" and "provoked conflict" (2003, 99); and see also Holloway 1999.

sociological approach is pursued to address the questions of how to fill in the "we" of the "society" that needs to define acceptable limits to reproductive and genetic intervention, maintain a shared definition of progress in this area, and produce credible regulation, governance, and policy.

UK PGD

The research informing this book took place in Britain during a period of dramatic technological convergence that may take future historians many years to unravel, even with the benefit of hindsight. In the 1980s, few people saw in the infancy of IVF a platform for the successor sciences to genomics. Indeed few people imagined IVF as anything other than a treatment for female infertility or, even more specifically, blocked fallopian tubes. However, the expansion of IVF throughout the 1980s and 1990s was driven not only by its popularity, its commercialism, or its soon-discovered capacities to treat a much wider range of infertility— such as male infertility and infertility of unknown origin—but by its potential to be used both to prevent genetic disease and to enable research on human embryos that might lead to stem cell derivation. As Robert Edwards (one of the codevelopers of IVF) has insisted throughout the technique's rapid clinical expansion to become, in less than two decades, a routine procedure practiced worldwide, IVF was always seen as an experimental research method with enormous potential (R. Edwards 2004, 2005a, 2005b). That potential began to become much clearer in the 1990s, with the successful cloning of Dolly the sheep by Ian Wilmut at Scotland's Roslin Institute and, only two years later, with Jamie Thomson and John Gearhardt's successful derivation of pluripotent human embryonic stem cells in the United States. These achievements were "crowned" by the publication of the first "complete" draft sequence of the human genome project in 2001. Britain's first successful hES cell line, WT3, was created by Susan Pickering, Peter Braude, and Stephen Minger at the London clinic where this study was based in 2003, one year following the United Kingdom's commissioning of a National Stem Cell bank (Pickering, Braude, et al. 2003).

As noted in the preface, it is a central argument of this book that the development of PGD played a pivotal role in relation to both cloning and stem cell research by providing the first "bridge" or merger between assisted conception, or more precisely IVF, and clinical genetics, thus establishing IVF as a platform technology not only for fertility interventions but for genetic diagnosis. In turn, the ability to "reprogram" embryonic development is at the heart of both cloning and stem cell technology. Increasingly, the genetic and the epigenetic, or the molecular and the morphological, have been recombined in accounts of situated genetic action *within* cellular environments, in some cases even reversing the powerful "one-way" coding function of DNA. As Ian Wilmut's experiments with somatic cell nuclear transfer demonstrated, the powerful egg cytoplasm "tells the DNA what to do."

Consequently, IVF, PGD, and research on human embryonic stem cells are now so inextricably intertwined that it is impossible to debate the social, ethical, or political implications of any one of them separately. They have all, in a sense, become "frames" for each other. A key focus that unites these technologies is the possibility of harnessing, *and combining*, mechanisms of genetic, embryonic, and cellular repair. In May of 2005, in London, at the opening session of the Sixth International Symposium on Preimplantation Genetics, James Watson and Robert Edwards shared the platform for a session titled "Back to Basics," in which the full implications of the merging of genomics and embryology could be seen in the potential not only to screen every child for abnormal polymorphisms but to discover the genetic and epigenetic causes of cancer. As the Nobel laureate and codiscoverer of the structure of the double helix began his talk in characteristically confrontational language, by asking "If we could make a better human baby by adding genes, why shouldn't we?" (Watson 2005) , so likewise did the codeveloper of IVF, Robert Edwards, decry the "disaster zone" of aneuploidy and point to the way forward for gene targeting in experimental embryology (R. Edwards 2005a).

A Social Science of Genomics

The past two decades have also seen the emergence of a new subdiscipline of literature on assisted conception and the new genetics, published since the mid-1980s, in which qualitative approaches, such as ethnographic fieldwork and interview material, comprise the main sources of data (see Thompson 2005 for a review). The effort to collect firsthand accounts of reproductive and genetic biomedicine has resulted in more than a dozen monographs (including those of G. Becker 2000; Bosk 1992; J. Edwards 2000; Finkler 2000; Franklin 1997; Inhorn 1994; Kahn 2000; Mitchell 2001; Rabinow 1999; Ragone 1994; Rapp 1999; Rothman 1986, 1994; Sandelowski 1993; Thompson 2005; and Throsby 2004). Already it is also possible to observe the rise of a number of important subfields including the anthropology of reproduction (Ginsburg and Rapp 1995; Franklin and Ragone 1998a, 1998b; Inhorn and Van Balen 2002; Strathern 1992b), the anthropology of genomics (Heath and Rabinow 1993; Goodman, Heath, and Lindee 2003, Rainbow 1996, 1997), and the anthropology of biomedicine (Brodwin 1999; Franklin and Lock 2003a,

2003b; Konrad 2005a, 2005b). Moreover, these emergent fields are not entirely "new" in that they build on well-established subdisciplines such as medical anthropology, the sociology of health and illness, and social studies of science and technology—as well as other disciplinary traditions, such as nursing, psychology, and of course medicine itself.[3] In the United Kingdom, where medical sociology is the largest of the subdisciplines of the British Sociological Association, reproductive and genetic technologies have been the subject of an increasing amount of research from as early as the 1980s (see in particular Homans 1985; McNeil, Varcoe, and Yearley 1990; and M. Stacey 1992), much of which was prompted by feminist concerns (see especially Spallone and Steinberg 1987; Stanworth 1987).[4]

A more recent surge of critical scholarly interest in social aspects of biomedicine has gained momentum in the context of what are often referred to as questions of "biosociality" (Rabinow 1992, 1997) and the rise of genetic technologies. According to Peter Conrad and Jonathan Gabe, coeditors of *Sociological Perspectives on the New Genetics*, published in 1999, we are at "the dawn of the genetic age" and "are also witnessing the dawn of the sociological study of the new genetics" (1, 3). As they note, most sociological work in this area began in the 1990s, and the number of sociologists working on such topics remains "small compared, for example, with those who work on problems like HIV/AIDS or stress and mental health" (1999, 3).[5]

Similarly, Alan Petersen and Robin Bunton, coauthors of *The New*

[3] There are far too many "precursor" studies of "genomics" to comment on here (but see further in chapter 2), in particular because of the prominence of questions about genealogy, shared reproductive substance, and kinship within the social sciences—where "blood relations" are a very old "version" of genetic ties (Franklin and Strathern 1992). However, it is equally true that what might explicitly be called the sociology or anthropology of biomedicine has not become as large a subfield—in terms of active researchers and funded studies—as might have been expected midcentury, when the turn toward molecular genetics began to raise classically anthropological questions about what Goodman, Heath, and Lindee call "genetic nature/culture" (2003); and see further Silverman 2003 and Weiner 1995.

[4] The role of early feminist critiques of reproductive and genetic technology is commonly overlooked and undervalued, and the enormous body of feminist scholarship on the social, ethical, political, and legal implications of new reproductive and genetic technologies has yet to receive due credit for anticipating, and very thoroughly examining, many of the issues that have since come to greater prominence in the context of the "new genetics" (see in particular Birke et al. 1990; Pfeffer and Woollett 1993; Corea 1985 and Rothman 1986).

[5] To make this claim is not to overlook the very important contributions British sociologists have made to the emergence of social studies of genomics, as is evidenced by a number of anthologies (Conrad and Gabe 1999; Glasner and Rothman 1998; Marteau and Richards 1996; Corrigan and Tutton 2004; Clarke and Parsons 1997) as well as monographs (Kerr 2004; Kerr and Shakespeare 2002; Turney 1998; Ettorre 2002). The point is simply that this recent flourishing of sociological interest in genomics is, in Britain as elsewhere, in its early stages. See also the British journal *New Genetics and Society*.

Genetics and the Public's Health, published in 2002, argue that although "the diverse imperatives" of new genetic knowledges and technologies are "changing how we think about our bodies, ourselves, and society," and are consequently "affecting the lives of everyone," we have only just begun to examine "the broader social and cultural contexts within which new genetic ideas arise, assume meaning, and are applied" (1–2). Countering the tendency to focus on novelty and transformation, Anne Kerr also points to the need to recognize that "genetic knowledge and the research and technologies with which it is associated mean many things to different people. They can be both mundane and unusual, helpful and oppressive. In order to understand these rich and dynamic processes we must move beyond the narrow categories of patient or lay expert, to look at genetics in the wider context of people's lives" (2004, 168).

Writing in the early 1990s in the introduction to one of the first major social science anthologies published in Britain concerning what she termed "the scientific revolution in reproduction," sociologist Meg Stacey drew attention to the neglect of social-scientific approaches in this area, and "the existence of a problem in the recognition of the social science role" (1992, 11). To illustrate, Stacey recounted her experience at a Department of Health consultation meeting on the new genetics in the 1980s, where she "failed to convince the assembled doctors and civil servants that there was an aspect to the developments [in genomics] which required fundamental social scientific analysis" (1992, 11), despite it being clear that "medical practitioners . . . are acutely aware of the immense responsibility using the new techniques places upon them" (1992, 13), and despite the overwhelming obviousness (to her) of the "social component of [genetic] research, particularly in relation to prenatal diagnosis" (1992, 11).

Importantly, it was not Stacey's intention, nor is it the aim of this book, to position social science as providing "answers" to the challenging questions and dilemmas that arise in the context of new reproductive and genetic technologies. To the contrary, as Stacey spells out clearly, the study of "social dimensions" must acknowledge their irreconcilable multiplicity: "To think of such technologies as having 'social' dimensions provides a way of thinking about the multiple nature of their impact. For if social life is a manifold and complex phenomenon, then the one perspective it affords is that of the complexity and interrelatedness of acts and effects. In terms of the disciplines, that apprehension is social science" (1992, 2).

This definition of "the social" as multipartite, relational, and complex, and of social science as a means of apprehending its "manifold" nature, positions the social scientist within a specific disciplinary form of attention that differs from those of other sciences in terms of both the methods it uses and its "findings," or results. Within social science, these methods

9

and results are not, as in science or medicine, orientated toward identifying the best, or most robust, "answer" to a particular question. In fact, the reverse is true: precisely by suspending the presumption that we can even know what a "right" answer would be, social scientists often seek to reveal the formative processes by which both questions and answers acquire specific patterns and shapes. In turn, it is possible both to look within familiar questions for more unfamiliar ones and thus to widen the scope of the possible answers available to be "tested" or applied.

"The existence of a problem in the recognition of the social science role," as Meg Stacey put it (1992, 11), has consequently both remained the same and changed amidst what Petersen and Bunton describe as an "outpouring" of social-scientific work on the new genetics (2002, 2) during the 1990s. While this body of work is substantial, and unprecedented, and has important continuities with many older branches of social-scientific inquiry, it is also, by definition, often exploratory and experimental, as it is at an early stage in its development. There is a challenge to this new field in terms of defining its "role," both because of its divergence from bioethics and because social science approaches may not easily be recognized as scientific, or even legitimately scholarly, from the perspective of the clinical and scientific criteria on which biomedical applications of genomic knowledge are primarily based.

Although *Born and Made* explores the social significance of new genetic knowledge and technologies from the specific perspective of PGD, the account it offers of reproductive and genetic transformation crisscrosses many terrains, moving from clinics to laboratories to kitchen tables and press conferences. At one level PGD is a technique, but it is also a choice, an experience, a threshold, a clinical specialism, a scientific achievement, and, as we shall see, a place from which a particular kind of uncertainty and ambivalence is generated alongside confidence in its refinement, expansion, and success. Apart from its immediate uses, PGD is also a symbolic technology for many people who never come into any direct contact with it and who may not have any specific idea of what it involves, because it is the technique referred to in debates about "designer babies"—a term that has become central to wider public debate for reasons this book explores in some depth. The intensely contested meanings of new kinds of genetic choice, technology, knowledge, and governance are the subjects of this book, which presents the results of a two-year ethnographic study conducted in Britain during a period of considerable public, scientific, clinical, and legislative controversy over the use of PGD (2001 and 2002).[6]

---

[6] Although the study on which this book is based was officially funded by the ESRC from February 2001 to August 2002, the period covered in this book is somewhat larger, roughly comprising late 2000 to early 2005.

10

The resulting account of PGD tries to avoid some of the pitfalls identified by Anne Kerr in her insightful review of the early sociological literature on "genetics and society," for example the overemphasis on genetic choice and decision making, and the novelty of new "genetic identities," to the neglect of the often more mundane and ordinary aspects of the "new genetics" (Kerr 2004). While these topics are central to the analysis presented here, this book has been written with the aim of moving beyond many of the stock characterizations that limit the terms of engagement with the difficult issues raised by new reproductive and genetic technologies such as PGD. While it is common enough to encounter the stereotypes of the ignorant lay public, the overambitious scientist, the patronizing clinician, the desperate would-be parents, runaway technology, or the toothless bureaucratic watchdog, these are often misleading and superficial depictions. Taking the trouble to look beyond these stock generalizations reveals a great deal more serious analytical thought, particularly among those closest to these new technologies, and reveals as well the often deeply paradoxical conditions they inhabit.

Whose Rights and Whose Wrongs?

Another limitation to contemporary debate about techniques such as PGD, both within and outside medical scientific circles, is that the debate is so often primarily judgmental: Is this the best technique? Is the technique morally right or wrong? Do the outcomes minimize harm for all involved? Such questions, like those concerning safety, are obviously essential and logically primary both in terms of maximizing care and in terms of moral responsibilities. However, these are clearly not the only questions that need to be asked, and neither is it at all obvious how, when, or even whether they can be answered—never mind by whom. As noted earlier, it is significant that many of the "solutions" provided by the Warnock Committee to questions such as "Should human life be inviolable?" were widely acknowledged to be "arbitrary," such as the fourteen-day rule on embryo research.[7] Similarly, there is no definitive "answer" in many clinical situations involving ambiguity or uncertainty, or even in cases that are completely straightforward medically but involve conflicts of interest, for example among closely related kin. At such times, the imperative of "right versus wrong" may itself be both inappropriate

[7] The "fourteen-day rule" became the limit point for permissible embryo research on the basis of the argument that it is at about two weeks' time that the "primitive streak" begins to become visible on the embryo (see further chapter 1, note 34). As the Warnock Committee noted in their report, "some precise decision must be taken, in order to allay public anxiety" (Warnock 1985, 65, see section 11.19).

and harmful. For these situations, in which there may be several incompatible "rights" and "wrongs," an approach that is primarily geared toward reaching a singular, definitive assessment obscures the importance of more equivocal, ambivalent, or hesitant "positions." In such situations the primacy accorded the need to reach the "right" answer forecloses the option of asking whether existing definitions of "answers" are suitable to such questions to begin with.[8] The extent to which the data collected for this study repeatedly indicated a pattern of emotional and ethical equivocation, or ambivalence, as a form of necessary moral contingency, suggests that this is a key area deserving further research, as well as being one that social-scientific approaches are particularly good at identifying (Kerr et al. 2000; Kerr and Shakespeare 2002; Kerr and Franklin forthcoming).

Toward a New Language of Social Description

Rather than responding to the inclination to provide advice, recommendations, answers, or guidance, which would involve reconciling several competing perspectives into a unified set of directions, this book addresses the oft-repeated question of how "society" addresses the challenges of new genetic technologies such as PGD, by allowing the various perspectives collected here to remain in an unresolved state. After all, there will not come a "point" at which all the questions raised by technological modification of human reproduction and heredity will be "solved." It is difficult to determine which of the two problems encapsulated by the "genetics and society" formulation are worse: the vagueness of its two halves, or the void of the "and" connecting them. The main aim here, then, is to offer an account of PGD that makes more explicit the contested terms on which it is negotiated, and asks what these divergent views can tell us. To do so, it is necessary to challenge some of the unhelpfully superficial depictions of PGD, such as the one reproduced in the first epigraph to this introduction, and to provide a more substantial critical model of some of the neglected elements, patterns, and forces at work in the socialization of genetic medicine. Above all it is necessary to begin to develop a language of social description that can identify some of the divergent, ambivalent, and often contradictory responses to the new genetics—as well as the gaps between these and more confident and secure certainties. Do these have a social form or pattern? What are their identifying characteristics, and how might we classify them?

---

[8] It should be noted that there are many traditions within ethics, medicine, and also science that acknowledge the inadequacy of a "right versus wrong," "good versus bad" model for decision making in the context of the new genetics, and indeed genetic counseling is founded on the principle of nonjudgmentalism.

Which factors, themes, or recurring motifs become more visible through sustained immersion in the world of PGD?

Repeated patterns already characterize this field, especially at the level of public debate, as we have seen. There are, for example, familiar generic forms in which the "social questions" of new reproductive and genetic technologies are often reproduced. The spatial and often geographical imagery of scientific knowledge "marching forward" into "uncharted territory" through "revolutionary breakthroughs" that will "transform the future of medicine" is so common as to have become one of the early twenty-first century's most hackneyed clichés. The specter of new genetic technologies developing at "an alarming rate," and in particular the idea of medical technology "racing ahead" of society, is so commonplace as to be almost taken for granted. The "horizon industries" of stem cell manufacture, regenerative medicine, tissue engineering, and cloning are frequently depicted through analogies to "a new dawn," "a new era," and "an altered landscape" of medical, scientific, and technological possibility, or as a "new frontier." Added to these commonplace analogies are the impact models of technology, as in "the social impact of PGD," which imply that, like some foreign astronomical body, a new technique has hurtled in from outer space and crash-landed into our living rooms.

This book tries to look both within and beyond the language of such analogies and their underlying premises, to ask what the knee-jerk repetition of them obscures and *what their alternatives might be*. Rather than depicting medicine and science as "ahead of," "beyond," or "outside" society, and pessimistically representing "the social" as perpetually lagging behind, science and society are depicted here as much more deeply intertwined. While it is not helpful to underestimate the radical novelty of many of the new techniques, choices, and dilemmas encountered in the context of new reproductive and genetic technologies, or the difficult issues they present, it is equally unhelpful to overprivilege technological innovation as if it were a force unto itself. Such a view limits the possible responses to the "genetics and society" question to mere reactions, precisely by building into that question the foregone conclusion that all we can ever do is respond to events that are always already out of control.

But what is the evidence on which such assumptions are based? After all, few are the scientists or clinicians in the fields of clinical genetics or reproductive biomedicine who do not encounter on a daily basis the immense moral and personal responsibilities of their work. Neither do patients involved in treatment programs take their moral and social implications lightly, as the second epigraph to this chapter demonstrates. In contrast to the "impact" and "racing ahead" models, evidence collected for this study repeatedly confirmed that *greater proximity to the actual decisions made during research and treatment involving genes, gametes, or*

13

embryos engenders more, not less, respect for the moral issues at stake in them. This show of respect exists in many different forms: for some scientists a primary moral obligation is to make as much progress as fast as possible, in order to develop the most effective techniques and eliminate the greatest amount of suffering. Even scientists, such as James Watson, who advocate quite extreme measures such as offering genetic assistance to people with low IQ (or Walter Gratzer, whose advocacy of "genetic privilege," reproduced as the first epigraph to this chapter, is offered in the "Afterword" to Watson's [2000] book), would not see their views as either amoral or irresponsible. Likewise, leading PGD experts, such as Chicago's Yuri Verlinsky, may be criticized for introducing "unethical" practices such as tissue-matching embryos to become donors for existing siblings suffering from a chronic disease, but it is notable that such innovations are often defended, both by their inventors and by others, as fulfilling a fundamental ethical obligation to help those in need (Kuliev and Verlinsky 2002).

In the Midst of Science, Science in Our Midst

Contrary to the image of genetic medicine "racing ahead" of society, or impacting upon it like some wayward meteor, the description here reveals an intensely social activity that is very much in our midst. Many of the choices and challenges described in this book may be "new" scientifically or medically, but the relationships involved—within families, between individuals and the medical profession, or between scientists and governmental regulatory authorities—are well established, and in some cases very old. Moreover, this social activity—the constant interactions that define and shape science and medicine as much as they do any other human activity—is exactly where to look to find out more about the value systems and judgments that are constantly being forged in the context of the myriad daily and difficult decisions to be made (from how to build a human embryonic cell line to whether or not to donate "spare" embryos to research) by the clinicians, scientists, health professionals, and patients involved in treatment. *Born and Made* tries to ask if there are ways of negotiating PGD "close up" that are relevant to the issues raised by the expansion of PGD as they are perceived from "far away." This has become a particularly important question in the context of reproductive biomedicine because of the simultaneously intimate and far-reaching consequences of "how far" reprogenetic intervention should be allowed to go, and because the relentless focus on the future often obscures the importance of the past, and the "now" of living with PGD.

Far from being spectacular or remote, this ethnographic account emphasizes more often what is very ordinary, familiar, emotional, and

14

recognizable about "the topsy-turvy world of PGD." Moreover, this emphasis also illustrates that it is not only in the hospital or the laboratory that social definitions of genes, genomes, and genetics are being tested and reworked. Genes surround us today as never before, having moved out of the laboratory and into popular culture, as well as literature, film, and everyday speech. We see visual images of genes in the form of DNA gels, markers, helixes, or chromosomes in advertising, cinema, cartoons, and on the evening news. Once a technical scientific term, "gene" has become an ordinary component of everyday conversation, used across a range of communications from the jocular to the juridical to the nonchalant. As Carlos Novas and Nikolas Rose argue, the gene has become a major component of our ideas of selfhood, identity, responsibility, and even citizenship in the twenty-first century (Novas and Rose 2000; Rose and Novas 2005; and see Rose 2006).

Genes are powerful, and consequently they are political—indeed they are a classic example of what historian Michel Foucault terms "biopower," in that they are inextricably linked to the idea of management of the population (in terms of health, for example), while also being regarded as inherent qualities of the individual. Significantly, they are both highly technical and commonsensical—like atoms, they are invisible, yet few would deny that they exist or are, indeed, the essence of "real" life. Evelyn Fox Keller argues that the gene is one of the most powerful scientific concepts ever created (2000), and as feminist sociologist and historian Barbara Duden has argued, "gene talk" has become ubiquitous, as are its constitutive, repetitive, and cumulative effects on all of us who are interpellated by the very idea of having genes, being genetic, or embodying a genetic identity (Duden and Samerski 2003). This power of "genetic information" to enroll us in *the certainty of itself* is what Dorothy Nelkin and Susan Lindee famously named "the DNA mystique" (1995).

Whether increasing amounts of genetic information are enabling new forms of choice and control over one's health and one's future, disempowering the worried-well with abstract risk formulations, or depriving us of our humanity, as Duden's German colleague Jürgen Habermas asserts, is difficult to determine empirically, for example in terms of quantitative measurement. The empirical evidence of how "gene talk" affects individuals has only begun to be systematically collected and analyzed (Condit 2004; Finkler 2000; Kerr et al. 1998a, 1998b, 1998c; Konrad 2000, 2003a, 2003b, 2005b; Nash 2004; Rapp 1999). While it is predictably the case that PGD can be used to reveal the ways in which new genetic "choices" create both benefits and burdens, the next step is to ask, as Monica Konrad does in her ethnographic analysis of "predictive" genetics, how corresponding social responsibilities and obligations take shape in and through these, and other, dilemmas (2003b, 2005b). This is

15

yet another important arena in which the view that "society" can never keep up, and is always "behind" rapid technological innovation, is especially unhelpful—as Rayna Rapp has powerfully demonstrated in her account of the resocialization of "objective" genetic information among her female informants undergoing amniocentesis (1999), and Jeanette Edwards demonstrates in her account of how kinship is used to model ideas of blood and genes (2000, and see also Finkler 2000).

To point to the ways in which "gene talk" mediates and travels between the worlds of objective scientific fact and lived social relationships is neither to deny the "reality" of genes nor to "relativize" their connection. To the contrary, a major advantage of social-scientific description is that it is empirical, meaning evidence based, and thus "factual" while also being able to account for the ways in which different factual registers compete. This is a crucial feature of negotiating genetic information in a context such as PGD, where the "genetic fact" of cystic fibrosis, thalassemia, or a translocation cannot be separated from the equally important facts of existing relationships, family histories, individual identities, professional responsibilities, or public health priorities.

The suggestion that there is no such thing as strictly objective genetic information in the context of PGD could be seen as either a confusing or even a provocative statement, but it is nonetheless one of the most powerfully repeated themes to emerge from both the social-scientific and the psychological literature on genetic testing (Marteau and Richards 1996). As this book demonstrates repeatedly, the process of communicating genetic information in the context of PGD, where professionals come from widely divergent disciplinary backgrounds and patients, although highly motivated, rarely have any clinical or scientific training, is extremely labor-intensive. Yet, even here, where the accuracy and reliability of genetic information could not be of greater consequence, and the effort to eliminate any possible sources of error is relentlessly pursued, communicative outcomes of consultations are the result of painstaking accumulation of detail, repetition, clarification, qualification, and not uncommonly confrontation as well. This somewhat surprising feature of PGD thus adds yet another set of dimensions to the ongoing effort to understand "gene talk" in action, including all the odd and often paradoxical ways that *in order to be "perfectly clear," genes can never make only one kind of sense* (Franklin 2003c).

What Is a Gene?

The multiplicity of "how genes mean" in the context of PGD is not only an effect of PGD's having a social character or dimension—as if without that contaminating layer everything would be purity of reason, order, and good sense. Even in "strictly scientific" terms there is considerable scope for what "preimplantation" or "genetic diagnosis" may mean. The terms "gene," "genome," and "clone" are highly varied and imprecise, despite being "scientific" in origin (Keller 2000). A "gene" can refer to a coding function, a location on a chromosome, a nucleic acid sequence, or a unit of heredity. All these are very commonly used meanings of "gene," and there are many more. Similarly, a "clone" can be an offspring of a single parent, an offspring that is genetically identical to its parent, an offspring that is "nearly" genetically identical to its parent, or an offspring that is created by fusing two cells from two different animals using what has come to be known as the "Dolly technique." Some people, including scientific experts, describe identical twins as "clones." Others, including Ian Wilmut, who created Dolly, would argue that she was not truly a "clone," despite the fact that she is called, famously, the first "cloned" higher vertebrate, because technically she was the product of cell fusion (Wilmut, Campbell, and Tudge 2000). "Genome" is no more precise either. A "genome" can mean all the genes in a single individual, all the genes within a particular species, or simply the "draft sequence" of a majority of genes within a small group of people, which is what the human genome map depicted when its draft completion was announced in February of 2001.

However, as Alice in Wonderland knew, words often work best when they can mean many things. The same is true of "designer baby." This term is often understood to mean a child born through genetic selection to have specific, deliberately chosen traits. It is a symbolic term, which, like "test-tube baby," connotes a futuristic world of technologically assisted reproduction. Whereas "clone" has a strongly negative connotation, "designer baby" is more ambivalent: Why not choose the best for one's offspring? What could be more natural? Isn't that what we do already anyway? Or, in contrast, is it right to have so much choice? Is it immoral for some people to be conceived as the expression of others' desires? Will it lead, as Princeton biologist Lee Silver claims, to a world of elite "genrich" offspring, as in the film *Gattaca*?

As this book illustrates, none of these descriptions bear even a passing resemblance to the experience of undergoing PGD. In contrast, many of the couples who attempt it come to PGD as a last resort, having undergone painfully traumatic experiences watching their children die of terminal genetic disorders, or having grown up with the knowledge they are, or may be, affected by a late-onset genetic disease such as Huntington's chorea. Others have a lengthy history of repeated miscarriage resulting from rare chromosomal translocations. Far from seeking offspring with "genes for blond hair, blue eyes, an imposing stature, and perhaps resistance to heart disease," never mind "those that help make a great artist," as Walter Gratzer suggests (2000, 235), these parents, or would-be parents, simply want a child who will survive. And as the contrasting epigraphs to this chapter suggest, the view of what PGD involves from "far away" may be quite different from that encountered "up close."

"Designer Babies"?

In the same way that the expression "test-tube baby" was widely resented by patients and professionals involved in IVF because of its stigmatizing and alienating connotations, many couples who undergo PGD are critical of the term "designer baby," which is seen to trivialize, and fundamentally misrepresent, what the technique involves, who undertakes it, and why. Far from seeking perfection, many couples who opt for PGD are enacting a profound sense of obligation, drawn from the experience of watching a child of theirs die after a life of suffering (often at only a few months of age), to do everything in their power to prevent imposing that burden again. Often for such couples the option of initiating a "tentative pregnancy" (Rothman 1986, 1994) that is dependent on the results of an amniocentesis test is too traumatic. They would rather be sure from the outset they are not creating a life that will end in a premature death. Hence, far from seeking desired traits, they are fulfilling a painful and expensive sense of obligation to act responsibly. And in contrast to the "designer baby" image of elite and choosy parents "buying" the most conventionally desirable traits for their offspring, many PGD couples say they would be happy to have any kind of child at all, as long as it does not have to be born with its own inbuilt genetic guarantee of a painful and premature death.

It is undoubtedly relevant that the term "designer" signifies a surprising range of divergent meanings. To design is to make according to a plan. But a "designer" garment can be either a bespoke outfit tailored to an individual's specific needs or simply an off-the-rack designer-label item identical to all the others. Wearing "designer sunglasses" usually means that you paid a lot of money for them, they are fashionable, and they associate you with celebrity. Buying a "designer evening dress" could mean either that you went to an expensive designer boutique and had one made by a top fashion guru or that you bought a designer-label ball gown in the bargain basement of your local department store.

From this perspective, the use of the "designer" adjective for PGD is not surprising, given that its contradictory and often paradoxical meanings can be seen to express a commonly encountered ambivalence toward increasing technological assistance to the beginnings of human life. On the one hand, assisted conception and genetic diagnosis are celebrated as

18

means of overcoming obstacles to pregnancy, avoiding genetic disease, and offering greater reproductive choice and control. On the other hand, these very same new forms of choice and control are often criticized, and feared, as unnatural, immoral, or unsafe. Some people worry that reproduction is becoming increasingly medicalized, commercialized, and geneticized. Other people worry that too much interference in reproduction will either diminish the human condition spiritually or damage the human population genetically. Frequent idioms in debates about techniques such as PGD include the "slippery slope," the "thin end of the wedge," and the need for "a line to be drawn."[9]

The Problem of Limits

However, no matter how great a sense of conviction—widely shared among parliamentarians, policy makers, scientists, clinicians, and the "general public"—that there must be limits to scientific intervention into the beginning of life, it is never particularly clear where those limits should be set, or when they should be changed. Britain has clearer legislation in this area, and has had "strict regulation" for longer, than any country in the world—even if it is also among the most liberal and tolerant in terms of allowing quite radical scientific techniques. However, even in Britain, where violations of the Human Fertilisation and Embryology Act are punishable under criminal law, the rules are "broken," and the rules also change. This is inevitable, as there cannot be ongoing social consensus around the regulation of human reproduction and embryology in a context of dynamic social, scientific, and technological change without changes also occurring in regulatory practice.

Thus, the paradox faced by the UK regulatory agency, the Human Fertilisation and Embryology Authority (HFEA), is that it must maintain strict limits to the use of all the techniques it licenses, but it must also change and adjust these limits periodically, which, by definition, means periods of uncertainty about what they should be.[10] It is another primary argument of this book that regulation by this method, like the Warnock strategy that engendered it, requires what we might call "sociological thinking." In contrast to prominent IVF consultants such as Robert

[9] For a perceptive analysis of the image of a "line being drawn" in the context of reproductive biomedicine see Hartouni 1997, and for an insightful account of public debate on new reproductive and genetic technologies by one of the members of Britain's Warnock Committee, see Holloway 1999.

[10] The period covered by this study is one such period of uncertainty, which is one of the reasons it has been useful to document in detail the forms of public debate that occurred in Britain at this time.

Winston who argue that medical professional judgment is too often compromised by rigid HFEA requirements, as well as to critics of the HFEA who argue that it is inconsistent and haphazard, its supporters might argue that the HFEA works effectively *because it works sociologically* to find a pragmatic path to workable social consensus.

Moreover, the "paradox of limits" faced by the HFEA can be sociologically modeled, characterized, and analyzed in a manner that is both diagnostic and predictive. For example, many of the changes to HFEA procedure examined in this book occurred via what has become an established, and controversial, pattern of "desperate" single cases of exceptionalism. The case of Diane Blood is one of the most widely publicized examples of the kind of challenge the HFEA faces in attempting to maintain limits that are at once "strict" and socially acceptable. Diane Blood's husband, Stephen, died suddenly of meningitis in 1995. Acting without authority, one of his attending physicians offered to collect semen from his corpse. Under the HFEA, it is illegal to use or store gametes without the signed permission of their donor. Since Stephen Blood had not signed permission for his sperm to be collected or stored, Diane Blood was legally prohibited from using it.

Mrs. Blood, however, was a professional publicist, and organized an effective national campaign to acquire her husband's sperm. Public sympathy flowed readily and indignantly in her direction. A devout Catholic, a devoted mother, and a tragic widow, Diane Blood became a cause célèbre whose campaign garnered nationwide tabloid support. Eventually an obscure legal means of enabling Stephen Blood's sperm to be exported to Brussels was engineered, allowing Diane to be legally inseminated. Following the birth of two sons by this method, Joel and Liam, Diane Blood returned to court in 2002 to sue for the right to put Stephen's name on her son's birth certificates. The government minister for health, Alan Milburn, was fined twenty thousand pounds under a ruling that found in favor of Mrs. Blood under European Human Rights legislation (*R v HFEA Ex Parte Diane Blood*, 6 February 1997).

The verdict was described as "a triumph for commonsense" by the Labour MP for Birmingham Hall Green, Stephen McCabe, architect of a Private Member's Bill to establish the right of posthumously conceived offspring to have the name of their deceased father on their birth certificate. However, another version of common sense might be that it is not in the best interests of a child to be conceived posthumously—which might equally be considered repugnant and immoral by some.[11] In such a situation, disagreement surrounds not only the legal responsibility of the

---

[11] For a detailed anthropological account of the Blood case, and public debate surrounding it, see Simpson 2001, and for a comprehensive legal review see Morgan and Lee 1997.

HFEA to prioritize the best interests of any children whose conception they have licensed, but also what version of "common sense" this should be based on.

Whose Common Sense?

A starting point for this analysis, as for the Warnock Committee, is that the debate about PGD must be one in which several kinds of common sense coexist. To some, nothing could be more obvious than a parent's or couple's sense of obligation to prevent serious genetic harm to their offspring by the most advanced technical means available. To others, such a desire is eugenic and immoral. To yet another individual, PGD is simply impractical and costly, or too unlikely to succeed.

This lived complexity of PGD contrasts with the polarized characterizations the media often promote in "for and against" stagings of debate. Like Charis Thompson's study of reproductive biomedicine, the ethnographic component of this project revealed "reflective and reflexive participants in the generation of accounts of what is 'really going on'" (Thompson 2005, 16), such as Anne's comments reproduced as the second epigraph to this introduction, in which she clearly provides her own point of view but also situates this view in a wider context of public uncertainty and points toward its longer-term social consequences. Anne's understanding that PGD represents "a very grey area" underscores the difficulty of setting clear limits to its use, while her reference to the need to "draw the line" confirms that there must be some limits. Implicitly, Anne's comments also reflect the tension between the needs of individuals, such as herself and her family, and those of the wider society. At once positioning herself within the "we" of her relationship with her partner ("we've used PGD") and the "we" of society ("What conditions are we going to allow PGD to be used for?"), Anne deftly summarizes the difficulties and tensions that produce a particular kind of contemporary uncertainty and ambivalence this book seeks to chart. In doing so, Anne also offers an example of how she herself has navigated this dilemma, taking as her starting point that "obviously . . . we didn't want to have another child that was going to die." This kind of nuanced and careful thinking from the context of being "inside PGD" is arguably an important, and underrecognized, resource for coming to terms with the dilemmas it will continue to generate.

Some of the most interesting perspectives on PGD came from people who had changed their mind over time, reflected on that change, and applied it as a means of interpreting the range of viewpoints around them. Hence, for example, the anger some PGD patients felt about their initial genetic diagnosis often changed into relief, and even appreciation, once it became clear that this diagnostic information might increase their chances of carrying a healthy pregnancy to term. In turn, their experience of this change of perspective could help them understand their family members' resistance to treatment more effectively. And, even if they failed, many couples took comfort from having reached a different understanding of genetic disease through the process of PGD. Above all this study found PGD to be a site of *extreme ambivalences*.[12] This finding was supported by others, such as all the contradictory statements that nonetheless made perfect sense, for example when couples described PGD as their "only choice" but also described their other choices thoughtfully and comprehensively, or all the descriptions of PGD as "impossibly" difficult while it continued its steady expansion during the course of this study.

Together, these voices from people whose direct encounters with PGD have enabled them to discuss many of its challenges so articulately are offered here as carrying their own theoretical and comparative power, akin to what Peter Redfield, in his *Space in the Tropics*, calls "thinking through the world" (2000). As such, the material offered in this book suggests that the world of PGD and its people are generative of new kinds of thinking and language, and that indeed the debate about PGD has been particularly rich and insightful in the context of debates about regulation. In this sense, a sociological approach is not only a fruitful way of understanding PGD but has in fact already become its operative mode, most obviously through the Warnock Report, in its assumption that the British public is always multiple, that affect is as important as reason in social life, and that the practical ethics built into workable legislation must be able to change through deliberation over time.[13]

Like many of the medical scientific techniques that are its subjects, then, this book is itself experimental. Chapter 1 explores the question "What is PGD?" as a way of both introducing this technique in more detail and charting some important aspects of its history and distinctive importance

---

[12] Significantly, this finding is amplified by Kirstin Zeiler's doctoral research on PGD, for which she interviewed eighteen British, Italian, and Swedish gynecologists and geneticists. In her dissertation she describes "an ambiguous outlook" among her medical-professional interviewees toward PGD, and an "articulated ambivalence" on behalf of PGD patients—findings that are remarkably similar to those reported here (Zeiler 2004, 2005; and see further on the theme of ambivalence Franklin 2005b; Kerr and Cunningham-Burley 2000; and Kerr and Franklin forthcoming).

[13] Many thanks to the anonymous reader who helped articulate these points and also noted that "the UK's HFEA embeds . . . a perspectivalism that has marked social theory in the age of Haraway's 'situated knowledges' . . . reminding anthropologists and sociologists that their own histories of sociality have often already been indigenized" (Reader Report 3).

in the evolution of the governance of human fertilization and embryology in the United Kingdom. This chapter examines the emergence of PGD in the scientific literature, looking in particular at the influential work of one of the United Kingdom's most eminent scientists, Anne McLaren, to promote its development, and the unusual political context that surrounded passage of the Human Fertilisation and Embryology Act, to which the first clinical success of PGD, in June of 1990, proved a crucial tipping point.

Chapter 2 provides an account of the research methods used in this study, in part as a means of introducing these to a wider audience. From the point of view of what is emergent, or even primitive, about social-scientific methods in the area of reproductive biomedicine, there is a more pressing need to be explicit about exactly how the research was conducted—or why certain approaches were more successful than others. Interestingly, this is a standard feature of scientific writing and thus an example of the kind of "interliteracy" that developed over the course of writing this book. Such careful documentation adds to a cumulative and comparative record of methodology that is essential to rapidly developing areas of inquiry.

Chapters 3, 4, and 5 depict PGD from the perspective of the people closest to it, including PGD clinicians, genetic counselors, patients, and PGD coordinators. Chapter 3, "Getting to PGD," examines the many routes people take to get to PGD, and the diversity of reasons people seek it out. This chapter foregrounds the uncertainties PGD presents from the outset, to both patients and clinicians, as well as the hopes and expectations it generates. Looking at topics such as why people feel obliged to undertake PGD, the changing idea of biological relatedness, attitudes toward adoption, fostering, and disability, and the feeling of being "on the cutting edge," this chapter examines the "before" of PGD treatment.

Chapter 4 examines the "topsy-turvy world of PGD" from the point of view of going through an actual cycle. While this experience is never the same for any couple, or even for the same couple over different cycles, certain features of PGD stand out as its defining moments. Examining the temporality of PGD—the dramatic moments of activity punctuated by lengthy waiting periods—and its emotionality enables an account of what PGD feels like from a range of perspectives. This chapter continues to document the decisions and choices involved in PGD, and the paradoxes this form of treatment often makes explicit for both patients and clinicians alike. In this chapter, then, it is the "present tense" of a PGD cycle that is the central theme.

Chapter 5 examines how couples think about "moving on" from PGD, and the issues they face at the end of treatment, looking toward its aftermath. This is an important area that has received relatively little attention, while being in many ways one of the most important and lasting aspects of treatment—both for the minority of couples who succeed and for the majority who fail. In looking at the "afterward" of PGD, this chapter also looks back to its "before" and "during" to consider how they work together to define "the world of PGD."

Chapter 6 attempts to combine an account of "the world of PGD" with the issues and concerns outlined in the introduction and chapter 1, in particular addressing how issues of uncertainty, ambivalence, and trust may be related to the governance of reproductive biomedicine. In attempting to identify some of the most important overall themes and conclusions from this study, this chapter returns to the question of "sociological thinking," and to the place of studies such as this one within the broader context of public debate. The work of the British philosopher and parliamentarian Onora O'Neill is used to interpret some of the somewhat paradoxical findings of the preceding chapters, and to consider their implications for the future of PGD. These points in turn structure the conclusion, where the main points are summarized with a view to identifying key areas for future research.

Chapter 1
What Is PGD?

It's how you approach the word "choice" isn't it? [PGD] gives us the choice of healthy from unhealthy, as opposed to choosing, you know, a blonde or a brunette, or a boy or a girl. This is a choice out of necessity. Not for any other reason. . . . So I don't think we have a choice really.
— Zoe, PGD patient

3 May 2003. The glossy cover of the *Guardian Weekend* magazine features a provocative image of a sonogram of a fetus reading a volume of Proust to accompany an article by Bill McKibben warning of the dangers of the designer baby era.[1] The reconstructed scientific image challenges the viewer with the issue of genetic enhancement, by depicting a superintelligent baby in the womb. In provocative visual language it conveys the idea of a new era of genetic manipulation and made-to-order, "superior" offspring.

The image was created by artist David Newton, who altered the familiar scientific imagery of ultrasound by providing the fetus with reading material—in the form of a literary classic associated with particularly deep and challenging questions about the human condition. This image is also "literary" in its references to Mary Shelley's *Frankenstein*, Aldous Huxley's *Brave New World*, and films such as *The Boys from Brazil* or *Gattaca*, which now comprise a distinct artistic genre of technoanxiety about reprogenetic futures (Jackie Stacey 2005; Turney 1998; Van Dijck 1998). Like the figure of the clone, the designer baby has become an iconic signifier within this genre of the dilemmas and risks posed by new genetic technologies.

The cover illustration is captioned "Condemned to be superhuman: The terrifying truth facing tomorrow's babies," leaving little room for doubt that the designer baby is a threat to humanity. This popular view of the dangerous designer baby is not only a ubiquitous media shorthand for the scandal of genetic science going too far but is equally prominently figured in recent books by leading public intellectuals in North America

[1] The article was an extract from McKibben's book *Enough: The Dangers of Being Superhuman* on the eve of its British publication in June 2003.

FIGURE 1.1. In this provocative depiction of a "designer baby" created to accompany a feature by US author Bill McKibben criticizing new reproductive and genetic technologies as "racing ahead," artist David Newton has modified Josh Sher's ultrasound images to depict an enhanced, "superior" fetus.

and Europe, such as Francis Fukuyama[2] and Jürgen Habermas.[3] Both authors identify preimplantation genetic diagnosis (PGD) with "designer babies," which have become the embodiment of the widely publicized fears expressed by geneticist Lee Silver in *Remaking Eden* (1999) of a future in which there is *too much genetic choice*.[4] As Fukuyama writes: "Geneticist Lee Silver paints a future scenario in which a woman produces a hundred or so embryos, has them automatically analyzed for a genetic profile, and then with a few clicks of the mouse selects the one that not only lacks alleles for single gene disorders like cystic fibrosis, but also has enhanced characteristics, such as height, hair color, and intelligence" (2002, 75).

Following Silver, Fukuyama claims that "in the future it should be routinely possible for parents to have their embryos automatically screened for a wide variety of disorders, and those with the 'right' genes implanted in the mother's womb" (75). Based on such predictions, he warns of an inevitable collapse of the boundaries between genetic screening, diagnosis, and enhancement. Fusing the imagery of information technology ("a few clicks of the mouse") with that of "routine" genetic profiling, he imagines a scenario of embryos being "automatically analyzed" and "enhanced." This image, straight out of Aldous Huxley, is deliberately chosen to reinforce preexisting dystopian fears. As Fukuyama explains:

> The aim of [*Our Posthuman Future*] is to argue that *Aldous Huxley was right*, that the most significant threat posed by contemporary biotechnology is the possibility that it will alter human nature and thereby move us into a "posthuman" stage of history. This is important, I will argue, because human nature exists, is a meaningful concept, and has provided a stable continuity to our experience as a species. It is, conjointly with religion, what defines our most basic values. Human nature shapes and constrains the possible kinds of political regimes, so a technology powerful enough to reshape what we are will have possibly malign consequences for liberal democracy and the nature of politics itself (7, emphasis added)

[2] Francis Fukuyama is professor of political economy at Johns Hopkins University, and a former member of the National Bioethics Advisory Council appointed by George Bush and chaired by Leon Kass.

[3] Jürgen Habermas is a philosopher who works in the tradition of critical theory; he initially delivered his thoughts on designer babies in response to a lecture by Peter Sloterdijk (2000) titled "The Human Zoo." The controversy in Germany about these two lectures frequently invoked reference to World War II eugenic practices.

[4] Silver's book, translated into fifteen languages, "describes incredible new ways in which people will be able to reproduce and choose the genes they provide to their children," according to his website, https://web.princeton.edu/sites/lsilver. Lee Silver's course on biotechnology and public policy at Princeton University uses several films, including *Frankenstein*, *Gattaca*, *The Boys from Brazil*, and *The Stepford Wives*, to explore issues related to reproductive and genetic selection.

Importantly, it is PGD that provides the crucial interface between reproduction and genetics for Fukuyama, thus providing the threshold imagery of a new era, by offering parents "the first step toward . . . greater
control over the genetic make-up of their children," he writes (75).

Like Fukuyama, Jürgen Habermas warns that the future of human nature is imperiled by the kinds of genetic choices PGD enables. In *The Future of Human Nature*, he claims that "genetic manipulation could
change the self-understanding of the species in so fundamental a way
that the attack on modern conceptions of law and morality might at the
same time affect the inalienable normative foundations of societal integration" (2003, 26). In this passage, genetic "manipulation" is positioned as a threatening agent of change, a danger to humanity, and an
"attack" on society. This depiction of genetic manipulation as destructive of the "inalienable normative foundations of societal integration"
not only suggests a tension between "genetic manipulation" and the social order but obscures the question of where genetic manipulation
comes from to begin with, as if, indeed, it were an invasive, alien force.

This view of genetic manipulation as *a force unto itself*, hostile to social order and integration, is repeated in many of Fukuyama's and
Habermas's dire predictions about an "automated" genetic future for
the human species. As Habermas claims:

> *The deepest fear that people express about technology* is . . . that, in
> the end, biotechnology will cause us in some ways to lose our
> humanity—that is, some essential quality that has always
> underpinned our sense of who we are and where we are going,
> despite all of the evident changes that have taken place in the human
> condition throughout the course of history. Worse yet, we might
> make this change *without recognising that we had lost something of
> great value*. We might thus emerge on the other side of a great divide
> between human and posthuman history and not even see that the
> watershed had been breached because we lost sight of what that
> essence was. (101, emphasis added)

Here, again, "biotechnology" is attributed a sinister agency: it is in our
hands, but we *might not even recognize* its potential to change our very
nature. Both this elusive power of biotechnology, as a causal force unto
itself, and the character of human nature we might lose remain unspeci-
fied in what Arthur Frank has called the "neo-essentialist" antigenetics
arguments of Habermas and their "neo-evolutionary" counterparts in
Fukuyama (Frank 2003). Ironically it is *too much control*, and *too much
choice*, that will take away our ability to determine our own futures.

Writing in the *Guardian Weekend* in a similar vein, Bill McKibben
urges his readers to pause to consider "the terrifying truth" ahead of us,

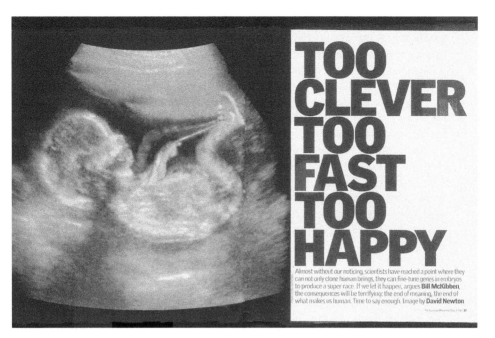

FIGURE 1.2. The possibility of physical or intellectual enhancement through the use of new reproductive and genetic technologies is one of the most common concerns expressed about the so-called "designer baby" technique.

and to decide to act—to draw the line against designer enhancements. Describing "where we are," he claims that "the genetic modification of humans is not only possible, it's coming fast: a mixture of technological progress and shifting mood means it could easily happen within the next few years. But we haven't done it yet. For the moment we remain, if barely, a fully human species. And so we have time to consider, to decide, to act." Like Fukuyama and Habermas, McKibben depicts a process that is *out of control*, and in need of restraint. He attributes the "terrifying" forward march of genetic modification to what he calls "a combination of technological progress and shifting mood" and argues it is time to call a halt. We are becoming, in the words of his article's title, "too clever, too fast, too happy"—a set of ominous predictions that are set in bold type opposite another of artist David Newton's "enhanced" fetal sonograms, this one tying the laces of its Nike trainers.

Newton's sportive designer baby in designer shoes signals the "end of meaning, the end of what makes us human." The theme of events acquiring a life of their own, and a barely perceptible momentum that allows them to slip past unnoticed (while moving "too fast"), is emphasized in

29

the article's subheading: "*Almost without our noticing*, scientists have reached a point where they can not only clone human beings, they can fine-tune genes in embryos to produce a super race" (McKibben 2003, 17, emphasis added). What is sinister, then, is not only the *nature* of the change at hand but the *process* by which it is occurring, without our consent, because we do not even notice. While scientists are busy producing "a super race," the general public remains unaware of the magnitude of the powers they have unwittingly allowed to threaten our species identity.

This popular but pessimistic view of scientists as powerful, untrustworthy, and socially marginal positions them as both ahead of, and outside or beyond, social norms, while they increasingly control forces almost too terrifying to contemplate. It is a view closely associated with the image of Dr. Frankenstein secretly creating a monster in his isolated laboratory, repeated as the motif in *The Boys from Brazil*, in which Nazi doctors plot to create a secret brotherhood of cloned Hitlers (Turney 1998). This *stock characterization* emphasizes a dramatic separation between scientists and society, and a conflict of interests between them. It exaggerates the remoteness of scientific knowledge and its disconnection from "ordinary life." While "we" are getting on with business as usual, "they" are designing made-to-order babies. Above all it is a view that emphasizes *secrecy*.

McKibben's scandalized and denunciatory rhetoric is closely aligned to that of Fukuyama and Habermas, then, not only through their joint call to halt the sinister, "unnoticed" forward march of cloning and genetic manipulation, but in their shared emphasis on the undesirability of genetic design and on the extent to which it is already routinized, automated, out of control.[5] The message from all three of these prominent authors is that the "designer" revolution in genetics is moving ahead too quickly, without our consent, and that it is not something we really want, because it will destroy who we really are.

The argument that designer babies are undesirable contrasts directly with the arguments of other commentators on the "genetic design" debate, such as Roger Gosden, a prominent medical scientist and IVF consultant, and the author of *Designing Babies*.[6] In contrast to the authors described above, Gosden argues that "far from being shocked that people

[5] In a lengthy critique of McKibben's article published in the *Guardian* science section (Life), MIT psychologist and author Steven Pinker (2003) refers to *Enough* as a "jeremiad" and claims it mistakes technological capacity for inevitability. (McKibben's first book, *The End of Nature*, 1989, equated the inevitability of the nuclear arms race with the discovery of atomic fission.)

[6] British reproductive biologist Roger Gosden, a specialist in ovarian and oocyte development, left the United Kingdom for McGill University in 2000 and became the scientific director of the Jones Institute for Reproductive Medicine in Norfolk, Virginia, in 2003.

should want to make 'designer babies,' we should expect them to" (1999, 37). As he points out:

There are now more than 300,000 IVF babies worldwide. Patients stand in line for treatment, regardless of the stress, discomfort, and risks and despite the fact that the success rate for treatment is seldom better than 1 in 5. The lengths to which some people will go to have the child they desperately want reveal more clearly than anything else *how powerfully we are driven by our desire to carry forward our own kind.* (1999, 26, emphasis added)

In Gosden's view, the desire for a "test-tube baby" is far from antisocial but is, rather, a vital expression of who we are. To couples such as those Gosden describes, IVF is seen to provide a means of satisfying the most essential function of humanity, "to carry forward our own kind." From this perspective, Robert Edwards and Patrick Steptoe, whose collaboration led to the birth of Louise Brown in 1978, were not engaged in an isolated, sinister, and antisocial form of experimentation with human life when they developed IVF but were responding to *a basic human desire familiar to everyone.* According to Gosden, the same applies to "designer babies": "We cannot deny the powerful drive within us to invest our very best in our children, and apply the benefits of discovery" (30). For this reason, he suggests, the designer baby option is not only inevitable—it is completely ordinary, predictable, and understandable.

A similar argument is made by bioethicist Gregory Stock,[7] who shares Gosden's view that far from being "some cadre of demonic researchers hidden away in a lab in Argentina trying to pick up where Hitler left off" (2002, 5), the scientists and clinicians who are laying the foundations for "the reshaping of genetics and biology" (2002, 5) are pursuing "mainstream research that virtually everyone supports," including treatments for infertility, which "is a source of deep pain for millions of couples" (2002, 5). In the conclusion to *Redesigning Humans*, Stock claims that rejection of new reproductive and genetic technologies is not only a misguided mission but a redundant one.

There is no way we can permanently forego these enhancement technologies if they prove robust and useful. Those who shun healthier constitutions and extended lifespans might hope to remain the way they are, linked to a human past they cherish. But future generations will not want to remain "natural" if that means living at the whim of advanced creatures to whom they would be little more than intriguing relics from an abandoned human past. (2002, 199)

[7] Gregory Stock is the director of the Program on Medicine, Technology, and Society at the UCLA School of Public Health.

Stock's advocacy of "greater germinal freedom" is thus the opposite of the fears shared by Habermas, McKibben, and Fukuyama in that he emphasizes the broad base of popular support for scientific intervention into reproduction, and the impossibility of separating biomedical innovation from treatment for conditions such as infertility.

Desires and Designs

But why are these arguments so focused on the idea of "design"? Why is "designer" equated with genetic "modification"? Technically, the only difference between PGD and IVF is that embryo selection is based on genetic information and morphology, instead of just morphology alone. And to be able to diagnose the presence or absence of a known, single, and specific mutation is not the same as modifying it. In addition to the famously imprecise concept of "human nature," a major area of confusion in the commentaries discussed above is the extent to which desire, demand, and design are constantly conflated in the depiction of genetic modification as "out of control." McKibben, Habermas, and Fukuyama view genetic "modification" as a sinister force that is racing ahead "unnoticed," driven by scientific desires, whereas Gosden, Stock, and other prominent scientists including James Watson argue that nothing could be more obvious than that parents would want to "apply the benefits of discovery" to provide "our very best for our children." *But what, exactly, is the agent of change, and what is the change itself?* Does the agency lie in the inexorable forward march of technology, as McKibben suggests? Or is it to be found in Gosden's biologically determined parental desires? McKibben's reference to "a mixture of scientific progress and shifting mood" suggests the dangers of too much genetic control *combined with* the threat of too much consumer demand, while Gosden views the desire to have children as almost synonymous with demanding the best for them, *including control over their design.* Is human nature the driving force behind technological innovation, as Mary Shelley's account of Dr. Frankenstein's desires to design and build himself a companion warns us? Or is human nature itself at risk of being eliminated by the "automatic" and unstoppable progress of science and technology? Is human nature the origin or the object of the "invisible hand" of genetic manipulation? Habermas warns that "the deepest fear that people express [is that] biotechnology will cause us . . . to lose our humanity" (2003, 101). But such warnings are confusing: Is something apart from us taking our humanity away from us? Or is this something we are doing to ourselves?[8]

---

[8] One might also want to ask on what basis Habermas assumes he can generalize about these fears, and which people he is referring to.

Confusion about the relationships between parental desire, consumer demand, and biological design emerges as a consistent theme in high-profile debates about "designer babies," in which vague appeals to "human nature" predominate. Equally elusive is the word "design" itself, which, according to the *American Heritage Dictionary* (3rd ed., 1992, 506) has several contradictory meanings in relation to the idea of purpose or intent. As a transitive verb, "to design" is "to conceive or fashion in the mind," or "to formulate a plan." As an intransitive verb its definition includes "to make or execute plans," or "to have a goal or purpose in mind." A design is synonymous with both an imagined object ("a plan, a project") and a subjective state, "a reasoned purpose, an intent," or a "deliberate intention." The way "design" links subjective states to projected (or idealized) objects has a remarkable similarity to the definition of "desire," also a transitive verb and on the same page of the dictionary. Definitions of desire include "to wish or long for, to want," or "a request or petition" (506). To have designs on something is to desire it, and to desire something is to have a "deliberate intention" in the sense of wanting something, which is "to have a goal or purpose in mind." To desire and to design, then, are similar: both are defined as mental states characterized by purpose and intent, and are orientated toward futurity, promise, and hope.

The close overlap between the meanings of design and desire can further be seen in the various, prevalent, and opposing ways "designer goods" are associated with consumer demand.[9] As noted earlier, the definitive "designer" commodities are those produced by leading fashion houses for elite clientele, such as the wealthy international consumer market in Gucci hand luggage. However, the rapid imitation of such commodities,[10] and the emergence of a cut-price "designer" market in off-the-shelf items for nonelite consumers gives a second, opposed, meaning to the idea of a designer product. It is now routine to visit a "designer warehouse" where either cut-price or imitation "designer" goods are available as bargain commodities.

In contrast, the individual "designer" item, created by an individual designer, for an individual client or customer—such as Dolce and Gabbana's outfits for British style icon Victoria Beckham—is something else yet again. Does "designer," then, refer to a one-off, bespoke item, tailor-made for a specific individual, as in the designer outfits worn by royalty and celebrities? Or does the term "designer fashion" include all the products manufactured for the general consumer market by designer firms, such as Donna Karan T-shirts or Calvin Klein "designer jeans"? If

---

[9] In Germany an IVF child is referred to as a "wish baby."

[10] This form of imitation is also known as cloning, referring to (commercially) fraudulent reproduction.

33

FIGURE 1.3. Paradoxically, a "designer" item can mean either a one-off bespoke item created individually for a customer by a leading fashion house, or merely another off-the-shelf commodity in a line of fashion bearing a designer's name.

a designer item is supposed to be unique, how can it also be marketed as a copy or part of a batch? Perhaps what is distinctive, and especially lucrative, about the "designer" commodity phenomenon is the extent to which it seamlessly appears *to include and indeed to unite* all of these rather different—in fact opposite—market phenomena.

"Dawn of the Made to Order Baby"

On the front page of the 23 February 2002 *Daily Mail*, the lead headline announces Britain's first "made to order," or "designer," baby. The article below announces the granting of a license from the HFEA to the Centres for Assisted Reproduction at the Park Hospital in Nottingham to use PGD to select not only embryos unaffected by a specific genetic disease but also those bearing a tissue match to an existing child. The tissue-matched offspring would become potential cord blood donors (or "savior siblings"). Raj and Shahana Hashmi from Leeds, whose son Zain suffers from beta-thalassemia,[11] are described as having won "an emotional victory" by becoming the first couple in the United Kingdom legally to be allowed to attempt to produce a "savior sibling" for Zain, using PGD.[12] The first case in the world of this kind, that of the Nashes of Colorado in 2001, whose HLA-typed offspring Adam saved his sister Molly's life through a transfusion of his cord blood, led to pressures on the HFEA to license similar treatment in Britain.

To the right of the "made to order" headline is a picture of one of Britain's leading celebrities, Victoria Beckham, a former Spice Girl and wife of England's football captain David Beckham. Known as "Posh" (Spice) for her glamorous outfits, Victoria is pictured with a bulging bosom and a broad smile beneath the caption "And Posh has some news, too." The connection between the announcement of the Hashmis' legal victory in their attempt to conceive a "made to order" baby and the reference to Victoria Beckham's pregnancy is the idea of design. On the left is a "designer baby," while on the right is another one, expected by Britain's leading "designer couple."[13] Both kinds of "designer babies" are much desired children-to-be, who will be born to parents eager to

---

[11] Beta-thalassemia is the most familiar type of thalassemia, also known as thalassemia major, Mediterranean anemia, or Cooley's disease. Children with this disease are born homozygous for a gene compromising hemoglobin production, thus causing severe oxygen shortage in the blood and associated complications including decreased life span. Two parents who are heterozygous for this disease have a one in four chance of producing homozygous offspring, a pattern of transmission deciphered by the prominent American physician James Neel.

[12] The HFEA announced on 13 December 2001 its decision that tissue typing in conjunction with PGD was acceptable in principle, and later, on 22 February 2002, that a specific license had been granted to Simon Fishel for the Hashmis' treatment (HFEA 2002a).

[13] The media attention to "celebrity pregnancies," which has gathered momentum since Demi Moore's cover pose, naked and pregnant, for *Vanity Fair*, is also being invoked in this image, which is, like most media representations, polysemic, "multipurpose," and thus open to several interpretations (see Tyler 2001 for a critical account of celebrity pregnancy).

FIGURE 1.4. On the front page of the *Daily Mail* two contrasting images of "made-to-order" are juxtaposed in the form of a "designer baby" for the Hashmis and a pregnancy announcement from the UK's leading "designer couple"—David and Victoria Beckham.

give them the very best they can offer and, in the Hashmis' case, eager also to give their son Zain what they desire most of all, which is a cure for his debilitating illness.[14]

The contrast between the two lead stories is at once playful and obvious, tacit and "crude," and thus both self-evident and ambiguous. The way the idea of a "designer baby" is being used to link these two conception narratives is through a pun on the idea of "designer," by referencing two of its most common, but divergent, meanings. The pun is covert in the sense that neither the headline nor the caption refers directly to "designer babies." As is often the case in advertising and in tabloid journalism, particularly in Britain, the viewer is invited to interpret this designer contrast by correctly filling in the missing meanings, which, in this case, are two different senses of "designer baby."

Punning, often considered one of the most unsophisticated forms of humor, is widely employed by the infamous British tabloids as a means of celebrating their populism and "common touch." All the tabloids compete to become "the voice of the people" by appealing to their "gut reactions," while simultaneously positioning their readers as skeptical and discriminating judges of human character and world events. The typical *Daily Mail* form of double entendre, then, invites a canny, "knowing" interpretation of juxtaposed "common truths"—that the designer baby is at once a serious matter and a "joke." Like much of the joking and punning about Dolly the sheep, the humor of this juxtaposition works at several levels: on the one hand the Hashmi article is about a very serious matter of a child's life being saved but also the possibility of genetic manipulation entering a new era ("dawn"), while the fact of Victoria Beckham being pregnant would not be front-page coverage were it not for her celebrity as the female half of Britain's most (in)famous "designer couple." Since Victoria Beckham is herself, like the Spice Girls and even David Beckham, often depicted as something of a "joke," the pairing of her "news" with the headline account of the Hashmis' victory, on the front page of the *Daily Mail*, sends a message that is simultaneously disturbing and reassuring: new forms of genetic modification and "made to order" babies are being allowed in Britain, but "we" still have a sense of humor and can joke about them. A new era of genetic selection may be dawning, but we can still have a laugh, and to the extent that a pun is always also something of a joke about itself, and its teller, we can laugh about the fact that the last laugh will always still be ours, and is, in that sense, inalienable, much as our genetic constitution— which should be most "our own"—increasingly is not.

---

[14] David Beckham had recently received publicity for designing his own label of children's clothes for Britain's famous department chain of Marks and Spencer.

The punning and the humor around the "designer baby" issue in the British tabloid press is thus in many respects a kind of pantomime of the sober and bespectacled debates reviewed earlier about the future of human nature in the age of designer genes. In both contexts, there is clearly confusion and slippage in the meanings of "designer baby"—a degree of "play" attributable in part to the breadth of meanings associated with the idea of design as well as its overlap with both desire and demand. It is hard to derive a clear argument from the debates about human nature for the same reason it is not clear what, exactly, the "message" is from the *Daily Mail*'s obviously deliberate juxtaposition of two quite different designer baby stories.

What do these ambivalent examples tell us, and how should we interpret them? Is it not predictable that designer babies should elicit widespread curiosity, polarized debate, and media hype? Or is the very term "designer baby" part of a deliberate attempt to sensationalize, or scandalize, PGD by its opponents? How does analysis of such phenomena fit into the attempt to "characterize" public debate over PGD? Should the effort be to "correct" misleading media stereotypes? Or is the fact that "design" has become such a central signifier evidence of a more complex, overdetermined social phenomenon?

Whether the terms "design" or "designer" are deliberately obfuscating, simply a shorthand for PGD, indexical of a wider set of issues to do with genetic control, or part of the usual media hype (or some combination of all the above), these examples confirm that the designer baby has become a highly contested nexus of conflicting opinion, much of which is confused, contradictory, and ambivalent. Beginning with the more "highbrow" commentaries of philosophers, journalists, scientists, and bioethicists we saw a range of divergent certainties, from the belief that the designer baby will be our undoing, to the conviction that it is not in the least threatening but is rather as natural as the urge to parent itself. Within these accounts we saw a strong emphasis on the idea of "human nature," and a related tendency to posit science and technology as forces unto themselves—threatening to race ahead of society, to harm it, or to be utilized in the name of reproductive and genetic progress, as Gregory Stock suggests. Without speculating on the influence of such accounts, which is beyond the scope of the analysis here, it is possible to read them as *diagnostic* of three significant tendencies in public debate about designer babies—be it in Britain, in Germany, or in the United States. The British tabloid, or popular media, accounts discussed above also share these tendencies, which are (1) to depict PGD as *a mixture of desire and design*; (2) to position PGD as a *threshold technology* or an *interface* to an improved or degraded future; and (3) to express ambivalence, confusion, and equivocation about the "designer baby" technique in terms of its future consequences.

Thus, although use of the terms "designer" and "design" to refer to PGD is technically inaccurate (as no design or modification is involved), and may be described as misleading and harmful (as many PGD patients do), those terms are still important from a cultural or sociological point of view as "placeholders" for issues that may be difficult to explain or even articulate. In other words, it is precisely because the term "designer" is both so vague and so ubiquitous that it is worthy of further investigation.

PGD Histories

To better understand how PGD came to occupy such a pivotal position in contemporary debate about "designer" reproductive futures, it is useful to return to its beginnings. The remainder of this chapter offers an account of how PGD was developed scientifically in the United Kingdom, where, as we shall see, it also played an important and distinctive role in public debate and in the process of devising legislation. One of the striking features of this history is the role of PGD in *focusing* and *clarifying* public attitudes toward reproductive biomedicine, as challenging issues so often do. Consequently, one of the most important parts of the PGD story in Britain is the decisive role of this technique as early as 1985 in the elaborate process of devising legislation to govern "human fertilization and embryology"—a process that took more than a decade to complete, and which is ongoing. PGD's significance to this history derives in part from its transformation from being a scientific possibility into a clinical reality—or "birth"—during exactly the time period of legislative "gestation" of the Human Fertilisation and Embryology Act, namely 1984–1991. Another striking feature of the "PGD debate" in Britain that is explored in detail below is the unusually prominent role of scientists and clinicians, and in particular embryologists, in both parliamentary and public debate. To follow, then, is an account of "UK PGD" that revisits some of the questions raised by the commentators presented above: If PGD is either a threat to human nature or a benefit, how would we know? Who would decide? What would the criteria be? As the debate over PGD in Britain over the past twenty years demonstrates, these questions have produced an elaborate and unique example of public deliberation over "the facts of life." Above all, PGD may be viewed with ambivalence and uncertainty—especially in relation to the future—but, ironically, it has played the opposite role in the past, when it was hailed as "winning the vote" in a decisive victory that enabled passage of the first comprehensive legislation governing new reproductive and genetic technologies in 1990.

39

The birth of Louise Brown in 1978 was seen to create not only a new kind of reproductive choice but a legal vacuum surrounding its use, as well as an immediate practical imperative to produce regulation (Franklin 1992, 1997, 1999b; Mulkay 1997). A resolution to this legal and regulatory challenge was both lengthy in its evolution and comprehensive in its scope (Gunning 2000; Jackson 2001; Morgan and Lee 1991). Britain's renowned Human Fertilisation and Embryology Act of 1990 (HFE Act) emerged twelve years after the world's first test-tube baby was born in Oldham, Lancashire. Its enactment concluded an unprecedented process of public consultation and parliamentary negotiation. The Act remains the most extensive, substantial, and detailed legal framework ever created to regulate and govern what had previously been the legally uncharted ("virgin") territory of human fertilization and embryology.

Since 1990, the Act has been copied by countries all over the world and is widely seen as a unique, exemplary, and distinctively "British" achievement that continues to set the global "gold standard" for governance of the post-IVF reproductive "revolution," while reinforcing Britain's self-proclaimed role as the home of "the Mother of all Parliaments."[15]

However, IVF was not the only technique to play a leading role in the dramatic battle to establish governance over human fertilization and embryology: PGD played an equally influential role during the passage of the HFE Act, and its importance to the shape of contemporary regulation of, and debate on, reproductive biomedicine in Britain has steadily increased, even to the point of precipitating a wholesale review of the Act, its regulatory agency (the HFE Authority, HFEA), and the entire basis of reprogenetic governance in the United Kingdom.[16] Thus, despite the low numbers of actual PGD cycles/patients in the United Kingdom, which remain in the hundreds (dwarfed by the vast demand for IVF), PGD has, from the mid-1990s onward, become such a pivotal technique, linking

---

[15] For detailed accounts of the emergence of the Human Fertilisation and Embryology Act, see Gunning 2000; Jackson 2001; Morgan and Lee 1991. For social-scientific interpretations of the legislation, see Mulkay 1997; Franklin 1997, 1999a, 1999b; Spallone 1996; Steinberg 1997.

[16] Following a lengthy consultation and criticism of the HFEA for its "mishandling" of several of the PGD cases described in this book, and also driven by the need to reduce the number of "Quangos" such as the HFEA, a Parliamentary Select Committee chaired by MP Ian Gibson began in 2003 to prepare advice to the Department of Health about the possible need for greater parliamentary guidance on questions such as human embryonic stem cell generation and, further to this, the merging of the HFEA with a new Tissue Authority into a Reproduction and Tissue Authority (RAFT). The results of this process, which will undoubtedly repay careful future investigation, confirm one of the central arguments presented here, namely that, at least in the United Kingdom, PGD is associated with a high level of well-informed public debate and considerable regulation—not their absence.

Human Fertilisation and Embryology Act 1990

CHAPTER 37

LONDON: HMSO

£5·85 net

FIGURE 1.5. Britain's famous "HFEA" (Act) inaugurated the world's most comprehensive legislation on human reproduction and embryology through the establishment of a unique national licensing body—the HFEA (Authority).

IVF to cloning and human embryonic stem cell research, that it is, more than any other "issue," the source of questions driving changes to science, governance, and policy. For these reasons and others, the "UK PGD story" provides an essential background to the scientific potency, clinical urgency, and political volatility of the "designer baby" question.[17]

The successful development of IVF is most closely associated with the partnership of Robert Edwards, a developmental biologist trained in animal genetics and immunology in Edinburgh, and Patrick Steptoe, an obstetrician/gynecologist and surgeon practicing in Oldham, Lancashire, although in many ways the precursors and templates of the technique were well established in the British scientific imagination long before (Squier 1994) and are the offspring of the unique climate of experimental embryology in Britain, which expanded rapidly during the postwar period (Graham 2000).[18] Thus, while the efforts of Edwards and Steptoe were, like those of Watson and Crick, or Wilmut and Campbell, exceptional in terms of zeal, determination, and above all success, the outcome of their efforts had, for decades, existed as a well-established, but still somewhat speculative, prospect in much the same way that the idea of the designer baby or the human clone exists today. Part and parcel of a vibrant scientific environment in which experiments in mouse embryology yielded new insights into early mammalian development, while micromanipulation of rabbit and sheep egg cells enabled practical developments such as oocyte maturation to be refined and improved, IVF and PGD emerged from a distinctive period in British biological research in which the work of many scientists "behind the scenes" had set the stage for the turning points when these high-profile techniques "went clinical" to tumultuous media acclaim (see further in Franklin 2007).

Robert Edwards's work with mice at Edinburgh, followed by his continuing interest in mammalian egg cells while working in Bunny Austin's lab at Cambridge, had long predisposed him, and many others, to think of IVF in terms that extended far beyond infertility treatment (R. Edwards 2004, 2005b). One of his Cambridge PhD students, Richard Gardner, was among the first to develop sophisticated biopsy techniques on mammalian egg cells, eventually sexing rabbit oocytes by this method (Edwards and Gardner 1967).

---

[17] While it is beyond the scope of this book to provide a complete account of this history, such a project would clearly repay much more detailed investigation, as has also been suggested by a number of commentators including Mary Warnock.

[18] The partnership of Edwards and Steptoe, like that of Watson and Crick, is also notable for its "forgotten" female, Jean Purdy, who, like Rosalind Franklin, provided crucial medical, scientific, technical, and administrative support to the first clinical success of IVF.

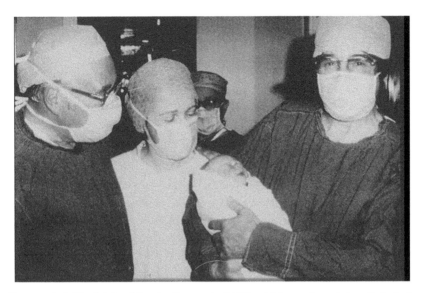

FIGURE 1.6. The birth of Louise Brown in 1978, which confirmed the clinical viability of in vitro fertilization, was itself the offspring of a particularly vibrant period in experimental embryology in the UK in the postwar period, which led also to the development of PGD and hES cell derivation.

Gardner showed that it was possible to transfer embryonic cells from one mouse blastocyst to another and produce viable offspring, and to sex rabbit embryos in vitro by embryo biopsy, assay of the biopsy for sex chromatin, and transfer of the surgically operated embryos back to a rabbit mother (Edwards and Gardner 1967; Gardner 1968). IVF, then, while widely perceived as a fertility treatment, was, from the beginning, clearly identified as a potential means of enabling embryo surgery and genetic diagnosis.[19] Since biopsied embryos appeared to be remarkably robust in all known research on mammals, the prospect of surgical removal of some cells from human embryos, their diagnosis to eliminate

[19] In his opening address to the Sixth International Symposium on Preimplantation Genetics in London in May 2005, Robert Edwards was at pains to emphasize the "roots" of both human embryonic stem cell derivation and preimplantation genetic diagnosis in the early work on IVF. Praising the "gorgeous inner cell masses" of cultured human embryos pre-IVF (2005a), he stressed, as in his publications on this theme in *Reproductive Biomedicine Online* (and for that matter in his coauthored book with Patrick Steptoe, *A Matter of Life*, (1980) chronicling the development of IVF), the inextricability of PGD, hES cell derivation, and even transgenesis at the "dawn" of IVF (R. Edwards 2004, 2005b; Edwards and Steptoe 1980).

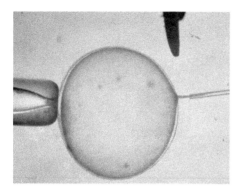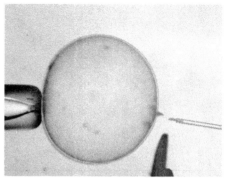

FIGURES 1.7. Richard Gardner's elegant biopsy of a rabbit egg trophectoderm provided an early prototype for the combination of micromanipulation with chromosomal analysis, thus laying the groundwork for one of the key components of PGD. (The left and right hand parts of the figure illustrate the before and after stages of this technique.)

specific genetic diseases, and the transfer of selected embryos to the mother to initiate a pregnancy appeared to be a viable strategy.

The Hope for Preimplantation Diagnosis

The prospect of developing successful preimplantation diagnosis (or PID, as it was initially known) was shared by many within the scientific community in Britain by the mid-1980s and was championed by some of its most prestigious figures, most notably Anne McLaren, one of the world's most prominent embryologists, who opened early doors to IVF through her work on embryo transfer using mouse models. The director of the MRC Mammalian Development Unit in London from 1974 to 1992, and the first woman to hold office in the Royal Society (from 1991 to 1996 as foreign secretary—she was elected a fellow in 1975), McLaren was also a member of the Warnock Committee and a key figure in the public debates around both IVF and PGD in the 1980s.

As a scientist with a keen interest in clinical applications, McLaren had been influenced in the 1980s by the work of two of her close colleagues at the University College Hospital (UCH) to view preimplantation diagnosis as an important research priority.[20] Bernadette Modell, a geneticist

[20] The technique that has become known as preimplantation genetic diagnosis (PGD) was initially described as preimplantation diagnosis (PID), referring to the possibility of any kind of diagnosis of the conceptus, be it chromosomal, genetic, or metabolic.

working closely with the Greek-Cypriot community trying to reduce the incidence of thalassemia,[21] and Marilyn Monk, a molecular biologist in McLaren's lab who had already begun to develop the sensitive single-cell molecular analysis that later provided the basis for the first successful PGD, both shared similar views.[22] For many parents in Modell's clinics, the experience of repeated pregnancy terminations became intolerable, not only because they typically took place in the second trimester, but because such terminations painfully compounded an accumulated sense of loss within the wider community. Some of these thalassemia patients— who were potential candidates for PGD—expressed their desire for an earlier means of diagnosis that would assure them of being able to begin a pregnancy free of the debilitating disease.

At Modell's instigation, she, McLaren, and Monk met to discuss the clinical possibility of preimplantation diagnosis, and to assess its technical feasibility. McLaren was inspired by the hope such a technique might offer to patients, but at the time felt the prospects were not good:

> We all sort of knew one another and Bernadette Modell knew that this work was going on and she asked Marilyn [Monk] and me and several other people to a little meeting—I think there were half a dozen of us in UCH [University College Hospital]—to discuss *what the possibilities might be of being able to diagnose at that very early stage*. And I think Marilyn was keener than Bernadette was actually, you know, being involved in those assays [on genes in single cells].[23] I felt that it would be possible eventually [but] I thought it would be a long time, because this was the 1980s, before PCR [polymerase chain reaction, a technology that will be discussed below]. (Interview, London, 16 June 2002, italics and footnote added)

[21] Bernadette Modell is professor of community genetics at the Royal Free and University College London Medical School, in the Department of Primary Care and Population Sciences, Whittington Hospital, London. As well as being one of the world's leading experts on beta-thalassemia in the community, she is also a public health advocate and holder of the UK Thalassaemia Patient Register.

[22] Professor Marilyn Monk is the former head of the Molecular Embryology Unit at the Institute for Child Health in London and a specialist on the activities of the X chromosome.

[23] Monk's contributions to PGD stem from her single-cell/single-gene molecular expertise, which enabled the first demonstrations of the feasibility of preimplantation diagnoses of genetic disease both in mouse model systems (Lesch-Nyhan (1987), beta-thalassemia (1988), sexing (1989), and in the polar body of the human embryo (sickle-cell disease, 1990). This expertise was considered so remarkable that she was invited to present her work to the Nobel secretariat in 1990. Monk continued to develop single-cell technologies and reduced the levels of contamination (error) rate to less than 5% (for over fifteen genes) in her academic research. This early development of single-cell technology provided crucial scientific confirmation of the clinical viability and potential of PGD.

CELL   SEQUENCE of INTEREST   AMPLIFY   ANALYSE

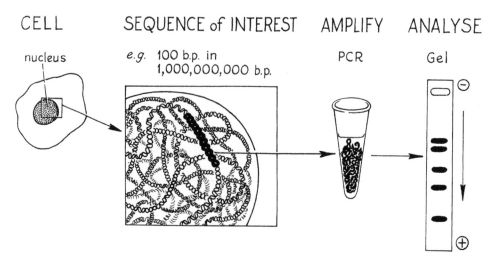

FIGURE 1.8. As this illustration demonstrates, a crucial component of PGD is the ability to amplify minute quantities of cellular material with sufficient reliability to produce correct analysis in at least 95% of cases (Holding and Monk 1989).

Despite her cautious outlook, McLaren began her 1985 review article, "Prenatal Diagnosis before Implantation: Opportunities and Problems," with an optimistic proposal, notably in the present tense: "One approach to prenatal diagnosis of hereditary disease is to remove a part of the conceptus for chromosomal or biochemical analysis, allowing the remainder to continue development if the results of the diagnostic procedure are favourable" (1985, 85). Such a diagnosis would primarily involve two novel technical procedures: removal of part of the conceptus, and its rapid chromosomal or genetic analysis. Citing, among others, the research by Steen Willadsen that later inspired the production of Dolly the sheep (Willadsen 1979, 1980), McLaren noted that the highly technically adept young Danish scientist had managed to produce normal lambs from a single cell removed from an eight-cell embryo, "rais[ing] the possibility that a human conceptus could be divided . . . or one or more cells could be removed . . . [and] used for diagnosis" (86), she noted.

The second task would be to analyze the contents of a single cell nucleus. Written before the discovery of polymerase chain reaction (PCR), through which DNA amplification and analysis was made much more readily feasible, McLaren's article noted that the options for such sensitive assays were not ideal. Marilyn Monk had shown it would be possible to measure the enzyme activity of specific genes on the X chromosome using "very small amounts of material from pre-implantation stages."[24] The alternative would be to create cell lines either from the preimplantation stages or the inner cell mass of cultured blastocysts—a laborious and time-consuming option. Thus, McLaren cautioned, "Even if diagnosis could be successfully carried out on such material, it is not certain that the remainder of the conceptus would develop beyond the implantation stage, or at least not often enough to make the procedure acceptable."

Although keen to explore every possible avenue—even including merging two halves of two unaffected embryos to make a chimeric offspring[25]—McLaren concluded her assessment of the prospects of PGD in the mid-1980s by stating:

> Although possibilities exist for preimplantation genetic diagnosis by means of embryo biopsy, much research on human material would need to be done to find out whether they were feasible, or whether they reduced unduly the chances of a successful pregnancy. Since they necessarily involve in vitro manipulation and subsequent embryo transfer, it seems unlikely that they would ever become the method of choice except for couples who were undertaking IVF for other reasons, or who had absolute objections to the post-implantation termination of pregnancy, combined with considerable determination and financial resources. For most couples, chorionic villus biopsy as early in the post-implantation period as is safe and practicable would seem to offer better prospects. (89)

This conclusion summarizes McLaren's mid-1980s position, which emphasizes the considerable technical obstacles to PGD, and in particular the problem of sufficiently amplifying tiny amounts of DNA for an assay sensitive and reliable enough to determine the character of an entire embryo from a single cell nucleus. However, one of the most important of these technical obstacles was already in the process of being dramatically redefined, through the work of the California molecular biologist Kary Mullis, whose ingenious method for amplifying DNA became the cornerstone of the biotechnology boom in the 1980s and won him a Nobel Prize in Chemistry in 1993. Ten years earlier, while Mullis

---

[24] The enzyme levels in females would be double those in of males, since both X chromosomes would be active at that stage and could thus be used to sex embryos (Monk and Kathuria 1977).

[25] McLaren cautions that both "halves" would have to be of the same sex "to avoid later problems with sexual differentiation" but admits that "the prospect of chimeric progeny might not appeal to patients" (1985, 87). She also notes that "if the technique of nuclear transplantation into enucleated fertilised eggs ever becomes simple and reliable enough to be used in clinical practice, implications for prenatal diagnosis would be considerable" (88).

PRENATAL DIAGNOSIS, VOL. 5, 85–90 (1985)

REVIEW

PRENATAL DIAGNOSIS BEFORE IMPLANTATION: OPPORTUNITIES AND PROBLEMS

ANNE MCLAREN

MRC Mammalian Development Unit, Wolfson House (University College London), 4 Stephenson Way, London NW1 2HE, U.K.

KEY WORDS Preimplantation diagnosis Embryonic biopsy *In vitro* fertilization

One approach to prenatal diagnosis of hereditary disease is to remove a part of the conceptus for chromosomal or biochemical analysis, allowing the remainder to continue development if the results of the diagnostic procedure are favourable. In the second trimester, for example, fetal blood may be sampled with the aid of fetoscopy, or amniocentesis may be performed to recover amniotic cells. Because second trimester terminations are traumatic both for the patient and her family, and for the medical and nursing staff, methods of diagnosis that could be performed earlier in pregnancy would be greatly preferable. Several diseases can now be diagnosed at 8–12 weeks, on the basis of samples of chorionic villi, recovered usually through the cervix (WHO Report, 1983). If the embryo proves to be affected, the pregnancy can then be terminated before the end of the first trimester, at a stage when the women is less aware of the presence of her fetus, and her pregnancy is not yet apparent to others.

Would it be possible to carry out prenatal diagnosis at a still earlier stage, before implantation? This would entail manipulating the conceptus *in vitro* and then inserting it into the mother's uterus, since no way can be envisaged at present by which part of the preimplantation conceptus could be removed *in vivo*. If the diagnostic procedure took longer than the conceptus could safely be maintained in culture, it would need to be stored, e.g. by freezing (Whittingham, 1980; Trounson and Mohr, 1983), until the results of the tests were known.

Up to the onset of implantation the human conceptus is 'free-living' in the sense that it is not attached to the mother in any way, and the environment provided by the reproductive tract can be adequately mimicked by an *in vitro* culture system. Blastocysts can be recovered from the uterus by trans-cervical flushing (Croxatto *et al.*, 1972; Buster *et al.*, 1983); earlier cleavage stages are located in the Fallopian tube (Croxatto, 1974) from which they cannot be recovered by any simple procedure. Oocytes recovered from the ovary and fertilized *in vitro* would therefore need to be used if any manipulations were to be carried out during cleavage.

Animal experiments suggest two possible approaches to preimplantation embryonic biopsy, one during cleavage and the other at the blastocyst stage. As we shall see, neither is very likely to prove practicable for the human conceptus in the immediate future.

0197–3851/85/010085–06$01.00
© 1985 by John Wiley & Sons, Ltd.

Received 13 July 1984
Accepted 20 August 1984

FIGURE 1.9. In her 1985 assessment of the prospects of PGD, the distinguished embryologist and Warnock Committee member Anne McLaren emphasized the considerable technical obstacles to its success.

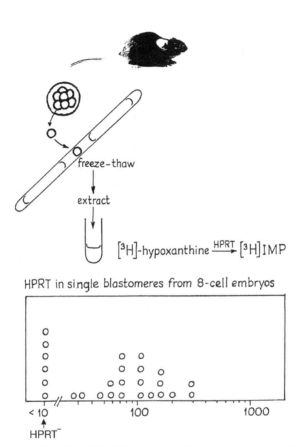

FIGURE 1.10. By the mid-1980s Marilyn Monk was among a small number of scientists convinced that highly accurate and reliable diagnostic analysis of the contents of a single cell could be accomplished using sensitive molecular assays (Monk et al. 1987).

was working for the Cetus Corporation, one of the largest American biotechnology firms, he developed a means of using the enzymes that copy DNA to produce millions of matching DNA strands from a single sequence of the total DNA.[26] This method and essentially solved the

[26] Anthropologist Paul Rabinow's account of Mullis's research, which is presented in *Making PCR* (1996), was one of the first major ethnographic studies of a specific scientific technique.

49

b. Thalassaemia $\beta Hb^+ / \beta Hb^\triangledown$

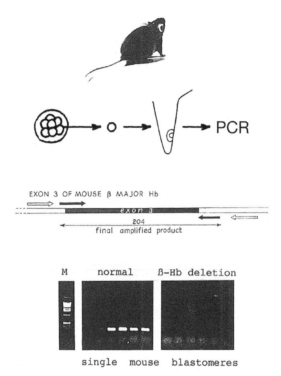

FIGURES 1.11 AND 1.12 (OPPOSITE PAGE). Beta-thalassemia and sickle-cell anemia were two of the major diseases for which preimplantation diagnostic techniques were successfully developed in research experiments using mouse and human embryos (Holding and Monk 1989; Monk and Holding 1990; Monk 2001).

problem of material from a single cell, previously seen to be one of the major obstacles to PGD.

In the meantime, the possibility of applying single-cell molecular analysis to the development of PGD had been continually advanced by Marilyn Monk, who provided the first demonstrations of preimplantation diagnosis of specific gene effects. During the early discussions of PID in the mid-1980s, Monk was convinced that it would be possible to demonstrate the clinical feasibility of PGD using single-cell assays for enzyme activities, which she had developed for her developmental studies on the inactivation of the X chromosome. Alan Handyside, who was then working at the Hammersmith Hospital as part of Robert Winston's

c. Sickle cell anaemia

Human

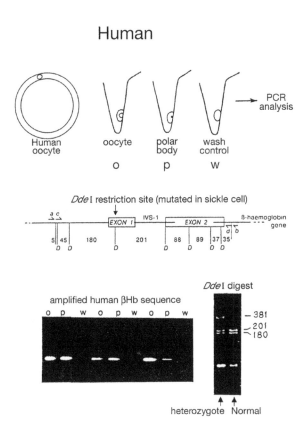

amplified human βHb sequence

Dde I digest

team, came to University College to collaborate with Monk on the biopsy of single cells from the embryos. Together they showed it was possible to diagnose the defective embryos in a mouse model for Lesch-Nyhan disease (Monk et al., 1987).

Following this, a key development was the application of the new technology of PCR (Mullis's method of amplifying DNA) to PGD. Monk and her colleague Cathy Holding began with the "water bath technique" for PCR and advanced to the use of the earliest PCR machines to detect specific gene mutations in single cells for PGD. Using a nested PCR technique, which increased the sensitivity and specificity of the molecular analysis, Monk showed that it was possible to diagnose beta-thalassemia in a single cell from eight-cell embryos from a mutant mouse lacking this gene (Holding and Monk 1989) and, with the collaboration of Peter

Braude, then working in assisted reproduction in Cambridge, established that it was possible to diagnose sickle-cell anemia in a single polar body of the human egg.

Returning to the theme of feasibility in an article titled "prospects for Prenatal Diagnosis during Preimplantation Human Development" in 1987, Anne McLaren, writing with her colleague Richard Penketh, could refer to important ongoing and unpublished work in the field. In a much lengthier and more detailed review article than McLaren had written in 1985, she and Penketh concluded two years later that the first human trials of PGD were imminent. The field, they claimed, had "advanced substantially" (Penketh and McLaren 1987, 761).[27] In addition to the much more robust technical case for the procedure's clinical viability, the 1987 review also includes many notable social, moral, and political references to the heated climate of debate, in Britain and worldwide, concerning the future of IVF and, in particular, embryo research. From a set of daunting technical problems that might never be adequately solved, PGD had become a clinical opportunity waiting to be undertaken, and a political question engendering heated debate in the halls of Westminster and beyond.[28]

Back at the Hammersmith Hospital Alan Handyside was able to translate this new and valuable technology into a clinical collaboration with Robert Winston that, following in the steps of Edwards and Steptoe, led to the first successful clinical application of PGD in 1990 (Handyside et al. 1990). As British science journalist Jack Challoner recounts in his popular history *The Baby Makers* (produced to accompany a Channel Four series by the same name aired in 1999):

> In 1987 Handyside was working at Hammersmith hospital with Robert Winston. In a frenetic two-year period, helped by teams at St Mary's Hospital Medical School in London and the Clinical Sciences Centre at Northwick Park in Middlesex, Handyside overcame the technical obstacles that lay before him. . . . Despite the difficulties, Handyside was eventually convinced the technique was ready to be

[27] For additional reviews of PGD and its development between 1985 and 1990, see Buster and Carson 1989; R. G. Edwards 1987, 1993; Edwards and Holland 1988; Edwards and Shulman 1993; Handyside et al. 1989; Monk 1988 1990a, 1990b; Penketh 1993; and Whittingham and Penketh 1987. For a comprehensive bibliography and review see Monk 1991.

[28] Numerous clinicians and scientists, including Peter Braude, Virginia Bolton, Martin Johnson, David Whittingham, Anne McLaren, Robert Winston, Robert Edwards, and their colleagues, played a prominent political role during parliamentary debate of the Human Fertilisation and Embryology Act from 1985 to 1990, during which they not only advised government committees on the drafting of the Act itself but also lobbied MPs and the Prime Minister, Margaret Thatcher, and wrote speeches for their supporters in both houses.

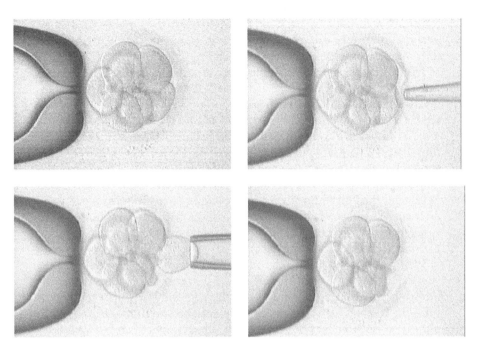

FIGURE 1.13. The highly precise handling of the embryo, which was enabled through the rapid improvement of micromanipulation during the 1990s, provides the surgical baseline of PGD.

tested on patients [in 1989]. A number of couples approached the team at Hammersmith, asking whether the work being done might benefit them. (1999, 89–90)

The Hammersmith team's success—the first clinical application of embryo biopsy, single-cell chromosomal diagnosis, and embryo transfer to achieve a preimplantation diagnosis by sexing—was the turning point in what had been a long process of cautious equivocation and ambivalence concerning the feasibility of PGD.

As noted in the introduction to this book, the Warnock Committee never reached a definitive moral conclusion about the status of the human embryo and instead chose to recommend a path of maximum social consensus, by which both IVF and potentially beneficial research on embryos would be permitted, but subject to strict regulation by a Licensing Authority backed up by Parliament and enforced through criminal law. However, although strict, the proposed regulations permitted not only all the embryological procedures necessary for IVF, such as embryo storage and freezing, but also numerous research procedures—including the fertilization of donated eggs for research (with the consent of donors) and the culture of embryos for research purposes for up to fourteen days—that, would be necessary for PGD.

The Warnock recommendations were controversial. The strongest reaction predictably came from right-to-life groups, who campaigned intensively in Britain during the 1980s against IVF and embryo research as well as abortion. Unsuccessful attempts to repeal the 1968 Abortion Act were introduced into Parliament by both Enoch Powell and David Alton. Powell's Unborn Children Protection Bill garnered considerable support in Parliament and would have prevailed with a parliamentary majority on its Third Reading had it not been talked out by opposition MPs. Had the Powell Bill succeeded, it would have brought an end to all embryo research in Britain, which prospect had, in turn, caused such concern within the medical scientific community that it began to organize more effective lobbying efforts.

Virginia Bolton, a Cambridge-based biologist working on fertilization, was among the first scientists to begin to pursue a much more active role lobbying parliamentarians following the near-miss defeat of the Powell Bill.[29] In Bolton's view, the future of essential research in the area of human reproduction was at stake, with direct implications for women's health, as her research concerned causes of IVF failure. Robert Edwards similarly considered the Powell Bill to be "one of the most serious infringements of individuals' rights ever to be placed before the Mother of Parliaments" (quoted in Challoner 1999, 72). In 1984, Bolton, Edwards, and other scientists, including Robert Winston, Peter Braude, and Anne McLaren, established an informal organization, PROGRESS, dedicated to increasing public awareness of the benefits of IVF and embryo research.[30] They were aided in their lobbying efforts by a number of passionately committed parliamentarians, including some, such as Peter Thurnham and Daffyd Wigley, whose children suffered or had died from inherited genetic disease.[31]

The stakes in the battle to preserve a liberal and progressive climate for embryo research in Britain had been laid down, and a dedicated

[29] Virginia Bolton is consultant embryologist in the Assisted Conception Unit at Guy's Hospital in London and a scientific inspector for the Human Fertilisation and Embryology Authority. She chaired the organization PROGRESS through its parliamentary campaign until the passage of the HFE Act in 1990.

[30] PROGRESS began as an informal grouping but later was established as an educational charity, the Progress Education Trust, or PET. This organization continues to play an active role in public debate, through events such as conferences and symposia as well as its online publication *BioNews* (http://www.progress.org.uk/News/Index.html). For a historical account of the idea of progress in relation to nature see McNeil 1987.

[31] Peter Thurnham and Daffyd Wigley were both parliamentarians with children who had suffered from genetic disease.

medical-scientific lobby set to work to win over the hearts and minds of the public as well as parliamentarians in the buildup to pending debate over its legislation.[32] At the center of this debate was a new political question—*how to govern treatment of the human embryo*. Not only protection of IVF, but the future of the United Kingdom's thriving climate of embryo research, were the concerns motivating the formation of PROGRESS—an organization that was in no small part founded to protect the future prospects of PGD.

The Prospects of PGD Revisited

It was thus amidst a passionate and highly charged climate of contestation over embryo research that Penketh and McLaren's 1987 review of the prospects of PGD was published in a leading clinical journal of obstetrics. Unusually for a scientific publication, the 1987 article foregrounded McLaren's position not only as a leading scientist in her field but also as a member of the Warnock Committee and PROGRESS, as well as the Voluntary Licensing Authority established in 1986 to provide interim guidance until the Warnock Report recommendations were implemented by government. McLaren's sensitivity to the prevailing social, legal, and ethical issues surrounding her field in Britain in the mid-1980s, as well as her primary expertise in reproductive biology during a period of unusual political activism for her profession, are clearly evident in the 1987 article, which vividly conveys the contours of mid-1980s debate over reproductive futures in Britain.[33] She occupied a unique position at the intersection of science, medicine, and policy at a period of significant debate about the future of embryo research, in which PGD played a central role as an example of what hoped-for progress could bring.

Her position had increased McLaren's sensitivity to the need to portray the actual human suffering that a technique such as PGD could alleviate. Consequently, the difficulties faced by patients such as those of Bernadette Modell, featured in the earlier review of PGD, are spelled out in almost sociological detail in the 1987 article, which begins with more

[32] British science journalist Jack Challoner's dramatic account of the debate over embryo research details numerous episodes of impassioned commentary from parents of children suffering from genetic disease, including the "unprecedented incident" that occurred during a debate over the Powell Bill, in which Wigley "became so annoyed that he thumped the wing of the Speaker's chair, which broke" (Challoner 1999, 73).

[33] These influences are also evident in McLaren's writings for a wider audience through the popular science magazine *New Scientist*, in which she advocates the development of PGD (1987).

than a page of persuasive testimony on their behalf. While prenatal diagnosis has brought "enormous advances," it has done so at a "high physical and emotional price," accompanied by "fear and anxiety," note Penketh and McLaren (1987, 747). Moreover, "some families are particularly unlucky," such as those who have had to undergo serial terminations of wanted pregnancies following a positive diagnosis of chronic and often terminal genetic disease. "The distress of such patients is a potent stimulus for the improvement of prenatal diagnostic methods," the authors note. A beta-thalassemia patient's response to the question of what would most improve her situation, described in personal communication from Bernadette Modell, is directly quoted: "To be able to start a pregnancy feeling committed to it, in the knowledge that it would not be affected" (747). Penketh and McLaren continue:

> The only prospect of starting a pregnancy which is known to be unaffected involves diagnosis of the conceptus prior to implantation. Discussions with high risk patients indicate that this approach would be acceptable to them, and this attitude is reinforced by public statements of need, for example the evidence presented to the Warnock Committee by the UK Thalassaemia Association stating that there was an urgent need for research into this problem." (748)

This unusually poignant, and pointed, opening passage to a highly technical scientific article published in a clinical journal of obstetrics and gynecology clearly reflects the animated social context of debate at the time. Several facts are being established here as a means of sociopolitical persuasion—that research into preimplantation human development has relevance not only to fertility but to genetic disease, that fewer pregnancies would be terminated if PGD could be developed, that the need for such research has been formally presented to the government by patients' associations, and that such research is, for all these reasons, "urgent."

In subsequent technical sections about embryonic development and possible means of testing embryos, the political climate is no less evident in the use of specific terminology, such as the new term "pre-embryo" to describe the developing conceptus up to fourteen days. This term was deliberately introduced by the pro–embryo research lobby in Britain during the 1980s in an attempt to establish greater scientific legitimacy for the fourteen-day limit introduced by Warnock, roughly based on the emergence of the so-called primitive streak, or the first visible sign of an individual's body plan, which gives rise to the mesoderm. By reserving the term "embryo" to refer to the emergence of a new individual body, as distinguished from the indeterminacy of the "pre-embryo," which is made up in part of potential fetal and placental tissue and thus not yet identifiable as "a potential individual" (since it is also a potential placenta, umbilicus, gestational sac, etc.), it could be claimed that research on "pre-embryos" occurred before the biological fact of individuation.[34]

Concluding their article with a section titled "Ethical Considerations—The Warnock Report," Penketh and McLaren note that preimplantation diagnosis using established and recently discovered techniques, as outlined in their account, could be undertaken well within the Warnock guidelines. They also note the Warnock Committee's recommendation that "the fertilization of oocytes donated to research with the consent of the donors should be allowed" before concluding that "it is therefore likely that the relevant animal experiments will soon be supplemented by studies on human material within the guidelines laid down by the Voluntary Licensing Authority" (1987, 761).

The British Way Forward

The case for preimplantation genetic diagnosis emerging from this article is staged and presented as a precise working model of the Warnock strategy. The opening references to human pain and suffering, the description of medical and scientific capacity to relieve this suffering, and the conclusion that implementation of this new technique falls within the established and agreed-upon guidelines for research together present a powerful and persuasive case for scientific progress on behalf of people in need, subject to strict regulation. The arguments by right-to-life campaigners that such research is degrading to the value of human life appear callous in the face of couples struggling with the uncertainty and trauma of repeated, and resented, pregnancy terminations. Indeed PGD could be used to reduce the need for such terminations. The arguments of disability activists that any kind of genetic selection demeans those who live with a disease being diagnosed by PGD is countered by the reference to patient associations representing sufferers of inherited diseases, as well as the families and communities most affected by them, who are actively petitioning the government for more research on PGD.

As an example of the kinds of moral and scientific positions most closely associated with PROGRESS, and as an instructive preview of the terms on which the advocates of PGD eventually triumphed and have since prevailed, Penketh and McLaren's arguments epitomize "the British way forward" in the field of reproductive biomedicine that became established in the mid- to late 1980s in line with the Warnock strategy. As

---

[34] For discussion of the "primitive streak" see Crowe 1990; Franklin 1997, 1999a, 1999b; Mulkay 1997; Spallone 1996.

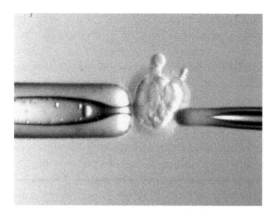

FIGURE 1.14. The possibility of performing preimplantation diagnosis through removal of the polar body (shown here as performed for research purposes) created a way around the anti-PGD argument that its use required the removal and destruction of embryonic cells (Monk and Holding 1990).

well as presenting a scientific overview of their field, and an assessment of its viability, the authors position their scientific arguments within a persuasive framework that emphasizes a social, human, and political context in which the causes of human suffering can be alleviated through publicly regulated and accountable scientific progress.

What is also noticeable about the role of PGD in this process of negotiating reproductive futures is its considerable "bridging capacity." It is at once a technique that holds out the promise of almost immediate future benefit but will still clearly require considerable ongoing scientific research to be applied safely and effectively—thus confirming the need for ongoing research. Successful PGD is set against an impressive backdrop of scientific achievement, much of it British based, ranging from Richard Gardner's elegant microsurgery on rabbit eggs at Cambridge, to Marilyn Monk's pioneering single-cell molecular techniques to detect specific gene mutations, to Steen Willadsen's prescient work on nuclear transplantation. PGD is thus revealed as a technique that embodies a distinguished scientific heritage and exemplifies the capacity of scientific progress to improve human futures. Part of the elegance and sophistication of PGD in medical-scientific terms is the battery of highly specialized, multidisciplinary knowledge and delicate, innovative, and precise techniques required to perform it successfully. This coupled with the personal, social, and clinical reasons why its practice would alleviate human suffering produces the classic "Warnock position" that urges slow, carefully regulated, and publicly accountable progress. With the added benefits owing to Marilyn Monk's successful polar body diagnosis of sickle-cell which demonstrated that PGD need not even involve any potential damage to the embryo; that its use would reduce the need for pregnancy terminations; and that its development had the support of leading patient groups for single-gene disorders, the political battle for PGD appeared to be on the cusp of success.

PGD: The Clinical Trial

In setting the stage for supplementary "studies on human material within the guidelines laid down by the Voluntary Licensing Authority," McLaren and Penketh somewhat understated the case. Alan Handyside and Robert Winston at Hammersmith were already prepared to approach their first clinical attempt at PGD by early 1989. They were successful in their first clinical trial, in which two of the three women treated became pregnant on the first cycle with embryos biopsied and tested for sex, using PCR to determine the presence of a Y chromosome (Handyside et al. 1990).[35] By implanting only female embryos, the possibility of a known risk of sex-linked disease was eliminated. It was the first successful use of PGD. Five days before the final Commons vote on the HFE Bill's Third Reading, in June of 1990, the Hammersmith team's success was announced in the journal *Nature*. Their subsequent press conference attracted headline coverage from national newspapers, including the *Daily Mail*. A mass lobby in Parliament featured speakers from genetic disease groups praising the technique and supporting its future development. The importance of embryo research had been shown to involve much more than infertility and IVF: the moral obligation to explore new avenues of preventing early childhood death and suffering from a range of conditions was linked directly to the prospects of human embryo research. These prospects of important future treatments of disease had a direct effect on public and parliamentary opinion. The HFE Bill was passed by a huge majority of 303 to 65 in the House of Commons—less than seven years after the Powell Bill, banning all research on human embryos, had won by a four to one majority in its Second Reading in the same chambers.

To speculate, in retrospect, that the timing of "the birth of PGD" was a direct result of the urgent desire of the medical and scientific community to ensure passage of the Human Fertilisation and Embryology Bill

[35] Handyside and Winston were successful in six of their first seven cases of sexing for X-linked disorders at Hammersmith. Although this is an excellent result, it could also be described as a high failure rate among those couples who reached the embryo transfer stage compared to the current rate of 1% to 3% that is considered acceptable by most clinics.

would be to exercise a degree of faith in the predictability of both medical-scientific and juridical innovation at odds with that held by any of their practitioners—much as it is a question that imposes itself upon the historically curious. There is no doubt, after all, that the political situation would have been on everyone's minds. But what *can* be observed about the order of events, in the Foucauldian sense of their "genealogical" unfolding, is that whereas the "birth" of IVF was seen to create a legal vacuum, the "birth" of PGD assisted in the opposite—by increasing public support for the Human Fertilisation and Embryology Bill to become part of British law. Here again, IVF and PGD pose interesting contrasts. Whereas PGD is currently (2005) dubbed the "designer" technique associated with "runaway" technology, while IVF is all but routine, in 1985 the opposite situation prevailed: it was IVF that had created a legal "vacuum" and PGD that helped to establish its social and political legitimacy. Moreover, far from being remote, secretive, and distant, scientists involved in the development of both IVF and PGD played a prominent role in the formative public and parliamentary debate about "human fertilisation and embryology"—a role that was sharply defined against a backdrop of the relief of human suffering that reproductive biomedicine either would or would not provide depending on the outcome of legislation.

The Authority Is Born

The Human Fertilisation and Embryology Authority began its work on 1 August 1991. With a detailed parliamentary Act to implement and a substantial legacy of public and private debate setting its agenda, as well as a preliminary infrastructure provided by the Voluntary and Interim Licensing Authorities, the fledgling HFE Authority was in a position to assert a strong regulatory regime from the outset. In the first decade of its work, its activities and staff expanded to encompass many new issues and techniques in concert with the expansion of reproductive biomedicine both commercially and scientifically. During the period 2001–2, this process of expansion acquired important new dimensions as the simultaneous, emotive, and high-profile issues of human embryonic stem cell research, "savior siblings," and cloning joining the Authority's always overstretched agenda. The challenges faced by the HFEA in relation to "designer babies" thus became enmeshed within a host of other issues, presenting an interesting case of national reprogenetic governance "in action."

As the regulatory body responsible for ongoing policy development and revision in a constantly evolving field of scientific and medical innovation, and as the executor of the parliamentary mandate to enforce the Human

Fertilisation and Embryology Act, the Authority has continued to pursue the Warnock strategy of trying to remain in step with public opinion. The Authority is thus in many respects the most obvious place to look for the criteria that determine "acceptable" and "unacceptable" technologies. As noted earlier, the Authority has the paradoxical responsibility to both set and enforce limits, and to change them. The mechanisms by which it does so, like so many aspects of the reproductive technology arena, did not exist until they were collectively invented. Since then, and still, they are constantly being refined and revised, while also being expanded and adapted to new circumstances (for a review of reproductive regulation in the United Kingdom see Jackson 2001, esp. 182–241).

On the whole, the Authority is respected and is seen to be run a tight ship, as well as being praised for generating 70 percent of its own income through funds collected from its licensees. Its primary function is as a licensing authority, overseeing all centers involved in the collection, storage, and use of gametes and embryos in the United Kingdom. To implement its licensing protocols, the Authority maintains a Code of Practice covering all aspects of infertility treatment and embryo research. This in turn is the basis for license applications, which are granted by the Authority on a case-by-case basis. This Code of Practice, in 2001 in its fifth edition, represents a deceptively simple set of criteria and protocols distilled from constant policy revision.

The Authority is headed by a Chief Executive, drawn from the Civil Service and charged with the overall duties of supervision, coordination with other government departments, and administration (the first chair of the Authority was Colin Campbell). The membership of the Authority (often referred to as the HFEA Authority) is composed of individuals appointed by the Secretary of State (the Authority had twenty-one members in 2002), more than half of whom must come from outside the medical and scientific professions, and they are chaired by the public figurehead of the HFEA and its most prominent spokesperson, who from 1995 until 2001 was Ruth Deech—the Authority's second chair (she was succeeded by Suzi Leather). At the time of this study, the membership and the Chief Executive were served by four departments: Resources and Corporate Development; Regulation; Policy and Communications; and Information Management. In addition to these, six committees reviewed various aspects of the HFEA's remit, including Scientific and Clinical Advances; Ethics and Law; Information Management; Auditing Procedures; Organization and Finance; and Regulation and Licensing.

In addition to producing a Code of Practice, operating a licensing system, conducting on-site inspections of clinics, and devising new protocols for emerging areas of research and treatment, it is the duty of the Authority to provide information to a number of constituencies, including

61

license holders and prospective licensees, people born as a result of treatment, donors, patients, prospective patients, government, the media, Parliament, and the general public. The Authority must maintain detailed records of its activities as well as the activities of the clinics it licenses, and this material must be made publicly available in a range of forms, from information videos to annual reports to the comprehensive HFEA website (www.hfea.gov.uk).[36]

The HFEA and its new chair were called upon to respond to a number of significant developments during 2001–2, the most prominent of which was the case of Raj and Shahana Hashmi mentioned earlier. Following the highly publicized treatment of Molly Nash at Yuri Verlinsky's clinic in Chicago, using cord blood donated from a "savior sibling," who was born following embryo selection for an HLA tissue match using PGD, the Hashmis began a campaign in Britain to be allowed to pursue similar treatment for their chronically ill son Zain. Like the Nashes, the Hashmis were an articulate, dedicated, and determined couple with a strong sense of obligation to do everything possible on behalf of their seriously ill child.

Coming from Leeds, they had approached one of the two clinics where fieldwork for this study had just begun in January of 2001, which was too early for the Leeds General Infirmary (LGI) to offer them treatment. Later, in 2002, the Hashmis came to an agreement with the Centres for Assisted Reproduction at the Park Hospital in Nottingham (CARE), directed by Simon Fishel, who agreed to work in collaboration with Verlinsky's Chicago clinic to treat the Hashmis in the United Kingdom. Before treatment could begin, however, it would be necessary to apply for a license to the Human Fertilisation and Embryology Authority, which requires all applications for PGD to be individually assessed. Since the Hashmis' case raised the additional issue of selection *for* a desired trait, not simply to exclude an unwanted one, their application raised novel issues that would require additional consultation.

The Hashmi case was considered by the HFEA in time-honored fashion, becoming the subject of a dedicated working group, as well as the Ethics and Law Committee, resulting in a proposal that was presented to the full membership for approval. In such cases, the Authority faces a difficult task. Its deliberating members must consult as widely as possible and be cautious, while not appearing to be mired in bureaucracy and dilatory in their efforts. Its aim, according to Debbie Jaggers, licensing manager at the Authority when she was interviewed amidst the Hashmi case deliberations, to find a way to respond from "pressure on both sides": "There's pressure for us not to move to eugenics and a slippery

---

[36] By 2001, when data collection for this study began, the HFEA had published five sequential Codes of Practice. The sixth was published in January 2004.

FIGURE 1.15. The expansion of PGD in the early twenty-first century has been led by cases of "savior siblings" such as that of the Hashmis in the UK.

slope and that side of things, and yet there's also the contrary pressure that there's a right to have [treatment], you know, if it's scientifically feasible, we should therefore have the right to have it" (HFEA interview 24 April 2001).

63

FIGURE 1.16. The Hashmi case raised controversy because it involved selection for an HLA tissue match as well as selection against a genetic disease—that is, not just "against" some genes but "for" other traits.

Negotiating these "contrary" pressures is both a procedural and a temporal process. Consultation with as wide a spectrum of views as possible is sought, as well as specialist expert advice where necessary. The views of other government departments and of relevant bodies such as patients' groups, medical charities, and research organizations such as the Wellcome Trust, the Nuffield Foundation, and the Medical Research Council also play an important role. As Debbie Jaggers describes the process of devising regulations, it is also about keeping pace with public opinion: "At the HFEA we want to move forward and be progressive but it has always got to be in step with the public's view, *and what people are comfortable with.* . . . So it's the safety, efficacy, and ethics, and moral issues [that we have to think about] and the public perception and comfortableness. I can't think of a better word, but you know, how comfortable they are with this" (HFEA interview 24 April 2001, emphasis added).

The domestic idiom of "comfortableness" both captures the highly affective quality of public opinion on matters concerning human fertilization and embryology,[37] and also conforms to the strategy of the Warnock Committee of maximizing public consensus by ensuring progress but exercising caution about "going too far." For the HFEA it is necessary both to be "forward and progressive" and to determine limits. The need for boundaries, for a line to be drawn, is a crucial part of the HFEA's role.

As Ginny Squiers from the HFEA put it: "People want this area of medicine regulated, and they see what goes on in America, they see what goes on in Italy, and they want some kind of boundary put on it." Keeping "in step" with public opinion, then, requires both the licensing of new treatments *and their prohibition.* This was exactly the strategy taken by the Authority toward the Hashmi case, and the others that inevitably followed.

For example, in the Hashmis' case embryo selection would have to involve both genetic diagnosis of beta-thalassemia as well as tissue matching to Zain for HLA compatibility. In the case of another couple, the Whitakers, for whom a license would be needed to use PGD to create a "savior sibling" for their child suffering from Diamond-Blackfan anemia (DBA), permission was refused.[38] Since DBA is not commonly an inherited disorder—but the result of a spontaneous mutation—selection for an HLA match alone would have no diagnostic and preventive function and would, therefore, set a new precedent of PGD being used solely for the purpose of creating offspring with specific desired traits. This was deemed unacceptable and in violation of the strict terms of the HFE Act.

The HFEA decision in the Whitaker case drew heavy criticism, both from members of the public who found the distinction between the Whitakers' and the Hashmis' cases baffling, and from authoritative bodies such as the British Medical Association (BMA), which stated: "As doctors we believe that where a technology exists that could help a dying

---

[37] To say "I'm not entirely comfortable with" something is often to express a feeling that cannot quite be put into words. The opposite of feeling "comfortable" is feeling "disturbed." A very brief way to summarize the "designer baby" controversy is that it is disturbing—a disturbance that causes a feeling of discomfort that is difficult to describe. The vagueness of the antinomy between "comfortable" and "disturbing" in the arena of reproductive and genetic innovation is one of the themes this book tries to explore in more depth.

[38] Human Fertilisation and Embryology Authority 2002a.

or seriously ill child, without involving major risks for others, then it can only be right that it is used for others. The welfare of the child born as a result of the treatment is of crucial importance. But in our view this is not incompatible with allowing selection of embryos on the basis of tissue type" (cited in Boseley 2001, 7).

Supporting the Authority from his position as a Labour Peer and government spokesperson for health in the House of Lords, IVF consultant and PGD specialist Lord Robert Winston asked: "Can you think of any other medical treatment which you would expect anybody to undergo without informed consent for somebody else's benefit?" Such a child, he argued, will carry "the spectre of being born for somebody else's benefit throughout his whole life" (quoted in Boseley 2001, 7).

In the case of the Whitakers, the Authority had exercised caution. Leaning in what Sara Nathan, an HFEA member and broadcast journalist, described as the direction of "restraint," the Authority drew a distinction that aimed to remain "in step" with public opinion (Boseley 2001). However, as these two cases demonstrated, "drawing a line" is especially difficult when it is also necessary to support innovative new technologies that cross over into uncharted ground.

Keeping "in step" with either public opinion or the views of medical professionals however, is also difficult when the public and professionals are themselves polarized over controversial issues such as the "savior sibling" question. While the HFEA decision to license the manufacture of the first UK human embryonic cell lines in February 2002 created little comment, the Hashmi decision drew extensive media coverage.

At a British Fertility Society (BFS) conference on the day following the announcement of the decision to grant a license to Nottingham to treat the Hashmis, for example, opinions among its members were clearly divided on the use of PGD for tissue typing as well as disease prevention.[39] Among comments from BFS members in favor of the Hashmis' treatment were the following arguments:

- A designer baby must be much more loved if its parents love their existing child so much they have gone to so much trouble to HLA-type a donor sibling.
- What is the difference if you use technology to get to a desired end more quickly?
- If the technology exists, how can you tell the parents they can't use it?
- If it is offered in Britain, it will be very carefully regulated; if not, people will travel to get it somewhere else.

[39] The British Fertility Society (BFS) is the leading British organization for medical professionals working in the field of reproductive biomedicine.

Opposing treatment, other clinicians argued:

- How would a child feel if he or she were brought into being as a donor, to serve someone else?
- What if more than one donation was needed? What if the sibling needs bone marrow? Or a kidney?
- Could the donor sibling really refuse consent to donate?
- Can society really afford to make such important decisions on the basis of single, desperate cases?

Significantly, the comments on both "sides" are far-ranging in their concerns. While some arguments rely on what are considered by some to be self-evident assumptions, such as the essential goodness of parental love and the superiority of the British legal system, others make reference to philosophical principles, such as the ends justifying the means, and the right to autonomy and self-determination. Still others invoke sociological questions about the form in which decisions are being taken: "Can society really afford to make such important decisions on the basis of single, desperate cases?" Others invoke technical uncertainties: what if another donation is needed? Although there may be two "sides," then, the concerns expressed on this issue cannot neatly be reduced simply to for and against. Moreover, many of these questions are unanswerable.

In reactions from PROGRESS and from right-to-life groups to the Hashmi decision, positions familiar from a long history of disagreement were polarized along predictable lines. Dr. Jess Buxton of PROGRESS Educational Trust (PET) wrote in an editorial in its online publication *BioNews* that the media hype about "designer babies" was a trivializing insult to the Hashmis:

A "designer baby" hit the headlines in the UK last week. . . . The story has provoked the usual cries of "designer baby" in some quarters, with fears that the ruling will open the doors to parents wanting babies with a shopping list of desired characteristics. . . . Those who say their decision will lead to parents "designing" their children are trivialising the Hashmis' situation, and that of other couples who have requested similar treatment. *PGD for tissue matching is simply a new technique to help couples have healthy children.* (*BioNews* 146, week 19/2/2002–25/2/2002, emphasis added)

In contrast, Peter Garrett, director of research for the United Kingdom's largest antiabortion charity, LIFE,[40] claimed in the *Daily Mirror* that

[40] LIFE is one of the leading pro-life groups in the United Kingdom, along with the Society for Protection of Unborn Children (SPUC) and the Commission on Reproductive Ethics (CORE).

FIGURE 1.17. Cases such as that of the Whitakers tested the ability of the licensing authority, the HFEA, to "keep in step" with the public's "comfort zone."

"designer baby" was all too accurate a term: "This case raises serious questions. While the term 'designer baby' is overused, it is all too appropriate in this case. The procedure will allow a child to be born just to supply bone marrow for an older brother. *Children are not commodities.* They should not be exploited" (*Mirror*, 23/2/2002, 11).

In cases such as those of the Hashmis or the Whitakers, it may be impossible to determine where the "comfort zone" of public or professional opinion begins or ends—and there is no obvious "middle ground." The pattern of public opinion being mobilized through high-profile individual cases that are mediagenic and controversial is so noticeable a feature of such cases as to be itself the cause of critical comment, as was evident among the BFS membership in their reference to policy being guide by "single, desperate cases."

As Mary Warnock made clear in the introduction to her committee's report, the field of human reproduction and embryology is highly emotive and is accompanied by deep moral sentiments, which are often acted out in "single desperate cases." These cases, as a number of BFS members astutely observed, become particularly volatile when medical treatments that are available, or practiced in other countries, are denied to people in desperate need of them. It is at such junctures, epitomized by the Diane Blood case discussed in the introduction to this book, that the duty of the HFEA to maintain strict limitations "on behalf of society" may come into direct and irreconcilable conflict with public sympathy for an individual or couple facing dire, and often tragic, circumstances.

An Objective Mix

In addition to seeking to remain "in step" with public opinion, the Authority relies on its established protocols and methods of decision making in difficult cases where a clear consensus cannot be reached. Through its extensive consultation structure its members aim to create a "mix" of views that guarantees the greatest "objectivity." As Chris O'Toole of the HFEA describes it: "The license committee would like the opinion of some peer reviewers, and that would be *a mix of people*. . . . We need a whole lot of expertise" (italics added). This would involve participation from clinical, scientific, legal, and philosophical experts, as well as religious views, the views of patients, and "lay" members of the public. As Debbie Jaggers put it succinctly: "The optimal situation would be *a total mixture*" of views. John Parsons, an inspector for the HFEA since its inception and a member of the Authority, describes the consultation process as one in which differences of opinion are deliberately sought out: "There's doctors, scientists, counsellors, nurses, administrators—all sorts of people with different agendas, so it brings an element of objectivity to the regulation which you might not get in other fields." This view of the HFEA licensing and consultation processes as highly mixed, and therefore more objective, again bears the hallmark of the Warnock strategy in its emphasis on *mobilizing social plurality as a means of creating publicly credible regulations* in what is widely recognized to be a highly contested and emotive field.

It is thus through debate, dialogue, deliberation, and consultation that the Authority performs the paradoxical task of *maintaining strict regulations while changing them*. The members must both keep "in step" with public opinion and, when this is divided, rely on being seen to be transparent, accountable, and judicious in their regulatory decision making. HFEA staff are civil servants at the center of a highly controversial field, and throughout numerous interviews and informal meetings during this study, which took place at a particularly intense period in the Authority's history, a striking feature was the absence of the intense emotionality that so often characterized other contexts of discussion of PGD. The "tone" of the HFEA was, in contrast, formal, precise, and rigorous. In a field of often passionate and polarized disagreement, the staff maintain not only a strict Code of Practice governing their licensing procedures but an equally strict informal code of practice governing their self-presentation as rational and bureaucratic.[41] As Debbie Jaggers describes the inevitable criticisms they encounter: "The tabloid press love the headlines from either the sort of things we have done or haven't done. . . . The public obviously raise concerns about a slippery slope when we allow something, but when we don't allow something its bureaucratic toothless watchdog, and yeah, it's a hard line." The "hard line" of the HFEA is "hard" both in the sense of knowing what line to take but also in the sense of being firm and robust. The staff of the Authority are at the center of what Debbie Jaggers calls the "hot pot" of conflicting opinion and divergent expertise, to which they must respond efficiently and pragmatically by determining the course of action most acceptable to the largest number of people within the guidelines that have been set by Parliament.[42]

In 2002 the Parliamentary Committee of Science and Technology called for reform of the HFEA Act to accelerate the process of keeping it in line with the public desire both for clear limits to be established in the field of reproductive biomedicine and for medical-scientific progress to be facilitated. These calls for reform were echoed by the incoming head of the HFEA, Suzi Leather, who suggested early on in her tenure that greater parliamentary guidance would be helpful on some of the issues facing the Authority. The period of this study thus coincided with a level of expansion in the scope of the Authority's responsibility that both confirmed its ability to respond to an increasingly wide range of issues—from cloning and stem cells to sex selection and savior siblings—and raised questions about its remit in the future.

Despite inevitable criticisms, many aspects of the Authority's performance have drawn praise, and its success can be seen in the steady

---

[41] Activities such as Friday afternoon desk tidying were part of the Authority's semiformal regimen of orderliness and routine.

[42] The "hot pot" is a particularly fitting analogy as it is a dish from Lancashire (Lancashire hot-pot), the county in which Louise Brown was born. The fact that Lancashire is the "birthplace" of the both the industrial revolution and the "reproductive revolution" has as one of its connecting threads the importance of sheep to this region—an animal that was essential both to the woollen industries that were the precursor for the industrial revolution and to the technique of IVF, to which ovine embryology was critical. Lamb is the main ingredient of hot-pot.

increase of participation from the constituencies most directly affected by its policies. Clinicians, for example, while not infrequently critical of aspects of the HFEA's performance, have on the whole welcomed its role, and have come to see the Authority as working on their behalf and to their benefit. The practice of on-site inspections, conducted by very few other countries, is credited with having created a climate of greater openness and exchange among the IVF community, which in turn is seen to exercise a much more effective means of self-regulation than in the more secretive and competitive context of the United States. In turn, the benefits of more robust accountability and transparency have been touted by the British government as one of the most attractive features of the R & D climate in the United Kingdom for both experimental research and corporate investment in the fields of reproductive biomedicine and stem cell research. The widespread perception of the HFEA as a "tightly run ship" enhances Britain's status as a country that can boast of having the world's most elaborate regulations governing reproductive biomedicine while also offering one of the most liberal climates for experimental treatments and research. These features of the United Kingdom's distinctive climate for innovative reprogenetic research—governed by established and comprehensive, but tolerant and progressive, legislation—have taken on additional significance as the fields of IVF and PGD begin to intersect with the emergent scientific research fields of cloning, stem cells, regenerative medicine, and tissue engineering. As is discussed further at the end of this book, the UK PGD story continues to twist and turn and has become, if anything, even more pivotal to the future of the medical-scientific community out of which it emerged, over a fifty-year period in which experimental embryology both "went clinical" and rewrote British law.

Conclusion

This chapter began by examining the arguments of some of the most prominent contributors to the debate about PGD, such as Francis Fukuyama, Jürgen Habermas, and Bill McKibben, all of whom emphasize the threat to human nature posed, and in some ways epitomized, by "the designer baby technique." A contrasting set of examples illustrated the opposing view, from Gregory Stock and Roger Gosden's popular accounts of PGD, in which human nature is seen to justify the "ordinariness" of the desire for technological progress, and to point toward the inevitability of the logic of parents using genetic information selectively on behalf of future children, much as Lee Silver has also done.

In contrast to the clarity and certainty of the positions taken in these

71

widely publicized, authoritative accounts, however, are the confusing and contradictory meanings of "human nature" they rely upon, and the ambiguous ideas of "design" or "designer" they evoke.

A similarly vague notion of "design" is evident in press coverage of the Hashmi case in Britain, which unfolded during 2002. In a typical representation of "designer babies" from the British tabloid press, a very different set of representational conventions were clearly utilized than in the *Guardian Weekend*, but a similarly ambiguous "message" about the designer baby question emerged, encompassing both the desirability of technological progress and its more threatening aspects—a task of "reckoning ambivalence" that is achieved in part through humor.

A close reading of the work of Anne McLaren, a leading figure in the British debate about preimplantation diagnosis, the story of "UK PGD," reveals an unusual combination of politics, medicine, science, and the law. Countering the image of scientists as remote and secretive, the history of PGD in Britain is one in which the opposite tendency proved prominent, as scientists left their labs to join clinicians and parliamentarians in pursuit of a shared commitment to the relief of human suffering through the legislative protection of scientific progress—a principle from which the chief lobbying organization, PROGRESS, derives its name.

Rather than science "racing ahead" of society, it is the sociality of science that emerges from this brief case history of "the birth of PGD" in the United Kingdom. Rather than being either a threat to the future of human nature or evidence of an attempt to redesign or "manipulate" the beginnings of human life, PGD emerges as the center point of a complex political history marked by unpredictable twists and turns, and variously aligned sets of interests that are as sociologically as they are technologically determined. The establishment of the HFEA demonstrates the capacity for science to be regulated through an extensive and far-reaching process of social negotiation and policy innovation that is ongoing. And against the view that the "designer baby" emerged without anyone noticing is the evidence of the crucial and explicit role of PGD in the success of the Human Fertilisation and Embryology Act, which would not be unrealistically linked to the first successful PGD cycles in the UK.

It is also possible to see more clearly from the history of the first ten years of the HFEA's operation as a regulator the importance of a pattern that has come to play a prominent role in its efforts to "keep in step" with public opinion. This is the pattern established by "single desperate cases" involving articulate and determined individuals and couples seeking access to treatments they are denied in the United Kingdom but which exist elsewhere. In response to some of these cases, such as that of the Hashmis, the existing limits to treatment have been extended, whereas in response to others, such as that of the Whitakers, the limits have been clarified and reinforced.

Above all, in response to the question "What is PGD?" it is possible to identify a number of distinct but overlapping analytical dimensions, from the technical to the legislative, and from the clinical to the philosophical. Across all these layers can be found instances of "sociological thinking" demonstrating how the Warnock strategy, outlined in the introduction, has repeatedly set the course for "the British way forward" in the field of reproductive biomedicine. As the case of PGD demonstrates, the Warnock strategy of promoting scientific progress subject to strict regulation has required constant public deliberation, both through the battles fought to secure passage of the HFEA Bill and in the way the Authority it established has implemented its regulatory duties and obligations. From this point of view, it is difficult to describe PGD as antisocial, or indeed to conceive of it from the perspective of so many stock elements of the "blue-eyed, blond-haired designer baby" debate.

Contrary to Habermas's claim that people's "deepest fear" about new reproductive and genetic medicine is that they will lose their humanity, the debate over PGD depicted in this chapter demonstrates a wide range of views about the desirability of PGD and about the appropriateness of the term "designer baby," and it shows the depth of feeling about both allowing and prohibiting treatment. Far from "two sides," the HFEA more accurately describes a "hot pot" of mixed opinion—a situation its members and staff have, again following Warnock, harnessed within their own working practice by aiming to maximize objectivity by maximizing the range of both expert and "lay" opinion they incorporate into their decision-making processes. In this way the HFEA has institutionalized the "sociological thinking" that formed the basis for the Warnock recommendations.

What finally, then, also emerges from this brief account of different versions of PGD and its future is the extent to which PGD in the United Kingdom, like stem cells in the United States, is associated with public debate and regulation, *not their absence*. Part of the value of examining the history of PGD in specific national contexts is the ability to learn through comparisons, such as those that might be made to any number of countries in which PGD has had a distinctive historical profile, for instance India, South Korea, Australia, Sweden, Denmark, Belgium, Israel, Russia, or Cyprus. This alone will not settle the ongoing, and often unanswerable, questions posed by new combinations of technological potential, parental desire, and the widely shared view of the need for limits to reproductive and genetic intervention. Neither would it be sufficient to point to the wide range of different strategies for regulating techniques such as PGD in the effort to suggest it is in fact society that is leading technology along a consensual path of greater relief of human suffering. The effort to present a "closer" and "thicker," or "experience near," account of how people navigate the issues surrounding the use of PGD offers the possibility that some of the terms on which PGD has already been debated might play a larger role in shaping its direction in the future. However, such an account will not eliminate the gaps between these views and others—which is fortunate, as disagreement may be one of the most important resources ensuring an ongoing debate.

Chapter 2
Studying PGD

Despite a number of classic studies, medical ethnographies as a group exhibit no distinctive theoretical or methodological features.
—Michael Bloor, "The Ethnography of Health and Medicine"

Analyzing the social life of PGD requires moving across widely disparate sites and materials, from media representations and scientific articles to interviews with policy makers and medical practitioners. It also requires a range of methodologies, from semistructured interviews to policy analysis and social history. Providing a methodological justification for using such a wide range of sources and methods, in what is intentionally a very open-ended and exploratory fashion, is not straightforward, and runs the risk of appearing rudderless, haphazard, journalistic, dilettantish, and unscholarly. On the other hand, it could be argued that the social science of biomedicine must be inventive to respond as creatively as possible to many of the topics it engages with. This chapter presents an account of the methods used in this study, in part as an effort to stimulate more discussion of this kind. Because of the increasing level of interest in the methodology of clinical ethnography, and because this is, as the epigraph to this chapter suggests, in some respects a surprisingly ill-defined topic, this chapter offers a retrospective "how [not] to" guide to the ethnography of biomedicine.

As George Marcus notes in his oft-cited discussion of "multi-sited ethnography" (1995), a common starting point is to "follow the object," for example by "tracking" PGD across several different sites while documenting the range of different contexts in which PGD figures prominently. The benefit of using these contexts comparatively is the ability to generate an interpretation of PGD that acknowledges its embeddedness in several different layers of social meaning and practice.[1] However, it is a notoriously challenging task to undertake multisited research of this kind, and there is a danger of becoming unfocused and losing precisely

[1] This method is very similar to that devised for the earlier ESRC study "Kinship in the Age of Assisted Conception," in which Franklin was a participant, although the present study, as will already be clear, makes wider reference to both cultural and science studies (see J. Edwards et al. 1993, 1999).

the specificity of immersion in a single place that makes ethnography successful.[2]

Like much anthropology of biomedicine, this book participates in a shift of focus away from many of the previous concerns of medical anthropology and the sociology of health and illness (Lock and Gordon 1988), and participates in the turn toward analyses of Western medical systems (Lindenbaum and Lock 1993). Earlier concerns of medical sociology, such as doctor-patient communication, the social construction of medical knowledge, and the critique of medicalization, have contributed to a widening interest in medicine as a dimension of sociality, not just a specific form of practice.[3] Central to these shifts is the rejection of many traditional dichotomies, in particular the distinction between illness and disease, and its attendant dichotomization of the biological body and the social world. The critique of these distinctions has, in many respects, run its course, and has been aided by biomedicine "itself" in the sense that many of the "biological facts" formerly assumed to be beyond any doubt have become more visibly contingent and uncertain (Franklin and Lock 2003a, 2003b). The "facts of life" are clearly prominent among the biological certainties once taken for granted but increasingly subject to dispute, often nowhere more vigorously than in strictly scientific terms. That fertilization must occur for an egg to develop is no longer either "true" or "self-evident." It turns out that biology's newfound plasticity is one of the wonders of our age.

Biology Remade

The increasingly explicit contingency of the biological, and the coincident rise in the social prominence of "health," have become increasingly publicly visible not only in terms of technical and scientific issues, but as intimate questions of identity and belonging (Lupton 1993; Novas and Rose 2000; Rose 2006). The biological sciences are explicitly viewed as a social, economic, and political issue to a degree never witnessed before (Franklin, Lury, and Stacey 2000; Goodman, Heath, and Lindee 2003).

---

[2] This book builds on Franklin's earlier study of IVF, an analysis of public, parliamentary, and media debate of IVF (in Britain in the 1980s) interspersed with "framings" of it from interviews and participant observation in an "effort to make visible the accumulated practices, assumptions and constraints which inform . . . contemporary discussion of new reproductive and genetic technologies" (Franklin 1997, 16).

[3] As Lindenbaum and Lock put it: "Often confronted with human affliction, suffering, and distress, fieldwork in medical anthropology challenges the traditional dichotomies of theory and practice, thought and action, objectivity and subjectivity" (1993, x).

This *socialization of the biological* is evident in the rapid proliferation of highly technical molecular themes and images, in particular of DNA, in Hollywood cinema, branding strategies, popular fiction, and mainstream advertising (Nelkin and Lindee 1995; Jackie Stacey 2005; Van Dijck 1998), as well as in high-profile disputes over everything from "Franken foods" to designer genes.

These are different issues from those faced by early sociologists of modern medicine, who sought to understand the production of the modern patient, the emergence of modern disease categories, and the questions of power, hierarchy, and control that came into existence through the "medicalization" of health. The questions posed by biomedicine as an object of social inquiry are generated not primarily in response to its authoritative claims, so much as by the *general social condition of which it is diagnostic*—namely one in which the biological dimensions of our existence *have become subject to an unprecedented level of deliberate choice*. Who will regulate these choices, and who will benefit from them? As leading disability theorist Tom Shakespeare observes, there is "no consensus on genetics" within the disability community, where genetics are seen as both potentially harmful and potentially liberating precisely because they "rebiologise disability" (Shakespeare 2003, 202; Rapp and Ginsburg 2002; and see Rapp 2003).

New Choices?

This is why it is important to document and analyze the many languages in which new genetic choices and decisions are currently being negotiated. These languages include those of clinicians, scientists, patients, policy makers, parliamentarians, journalists, academics, activists, and lobbyists—to name only a few. Since choices and decisions *are* being made, we might as well learn what we can about them by documenting and analyzing the languages, concepts, principles, emotions, and experiences that give them shape.

Such a view can rightly be criticized as overvaluing choice, by using this idiom to describe both the problem and the solution, when in fact many people have fewer choices as a result of what some have called "the fetishism of the gene" (Keller 2000), "genetic essentialism" (Nelkin and Lindee 1995), or "genetic discrimination" (Kerr and Shakespeare 2002). This is a fair and important criticism: "genetic choice" can be described as an Orwellian fiction as easily as a medical reality, and indeed this is the dilemma for which the "designer baby" has become a cultural signifier. It is equally true that it would be a mistake to overstate the importance of the new genetics, or their novelty, as to do so would uncritically contribute to the very "hype" that many say is harmfully misleading and disingenuous. Why overestimate or foreground the social issues raised by the new genetics when there are so many other pressing health matters affecting much larger numbers of people, and when in many respects the new genetics are not even new? Why should the high-profile "innovative" health technologies be the focus of social-scientific investigation, which is by definition limited and by some measures scarce, when only a tiny number of already privileged people will ever come into contact with them?

These are important questions that are frequently raised. As a perceptive speaker at a Cambridge seminar asked in response to a presentation of preliminary findings from this study in June 2002:

> I'm wondering if its worth it, or if it's even possible, to come up with other terms besides choice, like if there's any analytical language out there that gets us out of this phrase? I'm thinking of it without having to reuse the term because the term does carry so much ideological, political weight? So I'm wondering (a) if that's a problem and (b), if it *is* a problem, is there some other language and I've just been thinking about it?

This is an excellent example of the kind of methodological dilemma that is symptomatic of the difficulty of generating arguments from qualitative data—where the mere frequency of a word such as "choice" can make it appear to have a central importance. But what does "choice" mean in the context of PGD? As is demonstrated in the following chapters, this word increasingly appears to be a placeholder for questions that are much greater than the word "choice" can accommodate.

The issues raised by the idea of choice are also significant in that they are the place where "the personal" becomes political in ways that have not been fully charted, but are beginning to be addressed under the rubric of what anthropologist Paul Rabinow terms "biosociality" (1992) or what Rayna Rapp and Faye Ginsburg describe as the "biogenetic public sphere" (2002). As Marilyn Strathern has noted, the idiom of reproductive choice took on a new set of meanings as part of the "enterprise culture" created by Margaret Thatcher, which was based very much on the idea of individual and familial consumerism (1992b). Under Tony Blair as well, who like Thatcher has emphasized traditional familialism as a basis for citizenship, we can observe the overdetermined nature of consumer culture at the point of progeny—in terms of children as desirable acquisitions, desirable acquisitions for children, and the social desirability of acquiring children—all of which combine in the idiom of the "precious baby" familiar from IVF.

For all these reasons, this study had to be multisited because the social, cultural, and political significance of PGD, or reproductive medicine more broadly, cannot be confined to the clinic. Writing of Dolly the sheep in 2000, Anne McLaren proposed "to consider Dolly not as a sheep but as a node" connecting social and scientific "streams." PGD is similarly *a condensed node of social action*, or a nodal point of exchange between different "streams" of social action, for which a range of methods and concepts, as well as sites and locations, are required to give an account.

Participant-Observation

Historically, the role of the participant-observer was established in small-scale, non-Western societies and defined as a labor of translation, effected largely through selection, transcription, and re-presentation (Stocking 1983). Field notes were prepared on the basis of immersion within a particular social setting, and these were the basis for anthropological ethnographies, or monographs. Core analytic concepts, such as kinship, economy, or politics, enabled the anthropologist to provide a description of an unfamiliar way of life by collecting and recording observations (data), sifting through and sorting the data into themes, and then re-presenting an account of "a culture," or a cultural way of life, or a "society" by using recognizable categories, such as "economic life," "political structures," or "religious beliefs."

Early sociological ethnographies were focused on the ways of life associated with forms of labor and the professions, such as the famous ethnography of the medical community *Boys in White* (H. Becker 1961). Ethnomethodology, the technique most closely associated to Anselm Strauss, involved a "quest for immersive understanding" (Bloor 2001, 179) similar to that of anthropological ethnography, but was more concerned with the processes of socialization (such as those associated with the work of Irving Goffman) than the "culture" question associated with anthropology.

Largely because of the critique of colonialism (Asad 1973), ethnographic writing became the subject of extensive reevaluation during the 1980s, also influenced by the rise of critical theory and post-structuralism, which challenged many of the taken-for-granted certainties underlying universalistic models of "the family of man" (Clifford and Marcus 1986). In the space left by these questions, other models of observation, analysis, and comparison emerged. Writing of this shift, James Faubion claims that "if previously, culture was the fieldworker's question, it has since become his, or hers, to put into question" (2001, 39).

In the wake of the critique and expansion of the anthropological "field," contemporary anthropologists have developed a more diverse "tool kit" of concepts, theory, and methods, although many aspects of the labor of ethnography remain unchanged: it is still a process of collection, transcription, analysis, and re-presentation. Ethnography is still primarily a labor of translation, primarily achieved through reflection and writing. Participant-observation continues to be practiced across an increasingly wide range of social-scientific disciplines, but always primarily involves the combination of spending time in a particular community or setting and then "writing up" one's experiences.

Qualitative versus Quantitative Data

Personal experience has always played a central role in ethnography, and this has always taken two distinct forms—the one concerning the experiences of the people the participant-observer is working with, and the other concerning his or her own experiences. Since ethnographic encounters are inevitably highly personal, their "scientific" status has long been debated. This is not an issue particular to ethnography, since all sciences involve highly personal, and often intensely emotional, encounters: anyone who has read James Watson's account of the discovery of DNA, for example, will know that experimental science is highly personal and very passionate, regardless of how "objective" and "detached" its outcomes are (Watson 1968, 2000).

The situation of the ethnographer is different insofar as these emotional and personal relations are *more explicitly the focus of investigation*. Since social scientists are concerned with social action, social effects, and social relationality, their own relationships "in the field" are inevitably, and formally, both evidence and method. In other words, it is the ethnographer's own experiential participation in the social contexts he or she is investigating that provides *both the primary data he or she will analyze, and the means of doing so*.

Ethnography is one of the most highly qualitative methods in all of social science, and it is one of the discipline's lasting achievements to have "invented" this investigative technique, which has now been practiced for the better part of a century. One of the most distinctive features of ethnography is its *inversion of the traditional scientific method*, by making a hypothesis *an outcome* rather than a starting point. This is one way to describe the main difference between quantitative, or experimental, methods and qualitative ones, and deserves a brief consideration here.

Quantitative methods are designed to minimize bias by using an instrument or an apparatus to distance the scientist from his or her subject of inquiry. Hence, a questionnaire, for example, acts not only as a data collection device but as a *boundary* between the researcher and his or her subject. In contrast to an interview, in which the face-to-face personal interaction between the researcher and his or her subject can be seen to "contaminate" the data, questionnaires are considered "more neutral" and "less biased" because the questionnaire is "itself" an impersonal object. Questionnaires are consequently a highly effective means of producing quantifiable results that are considered "reliable" or "sound" insofar as they provide an "objective" measure of responses to an *instrument*, not a person. Depersonalization, in this model, is linked to both objectivity and reliability.

However, ethnographers cannot rely on a distinction between a personal encounter and an encounter with a neutral instrument, because they *are their own instruments* when they are conducting interviews, participant-observation, or fieldwork. It is the ethnographer's own involvement in a situation that *produces* hypotheses, rather than tests them. These hypotheses are also different from those routinely encountered in other kinds of scientific practice. For example, an experimental scientific hypothesis often concerns the existence of causal connections, whereas anthropological hypotheses often result from the observation of disconnections, or gaps. Ethnographically derived hypotheses often result from confusions, misapprehensions, misrecognitions, and surprises: a hallmark of high-quality ethnographic data is that it provides "completely unexpected" findings. Key ethnographic insights often come either from things that "make no sense" or, paradoxically, repeated statements that are "completely obvious."

Highly instrumentalized questionnaires, such as large-scale polling methods that use random sampling to maximize objectivity and representativeness, are a ubiquitous source of data about "public opinion." However, they are very limited in what they can reveal about "what people think" in part because they work best for questions about which people have already made up their minds. Polls and questionnaires are most effective when the questions are easy to answer. For example, a questionnaire that asks people whether they prefer hot dogs to hamburgers will produce a higher response rate than one that asks people about a difficult or controversial subject, such as whether or not they would eat genetically modified soy. Questionnaires rely on people *knowing what they think*, and being able to produce a simple answer to a simple question immediately. Before it is ever implemented, a good questionnaire is designed with a clear knowledge of exactly what it intends to find out.

Questionnaires work well to elicit responses about which respondents are confident and, above all, certain.[4]

The reverse is true of ethnography, which relies on the assumption that we may not know what the important questions are, or why, or how to ask them. Good ethnographic investigation thus often produces its most valuable findings *as questions rather than answers*. And ethnographic methods are especially useful for investigating areas of uncertainty, in which participants are unsure what they think, have ambivalent or contradictory attitudes, or change their minds. Whereas questionnaires work best when they strictly delimit the frame of responses within set parameters, ethnography does the reverse, by attempting to remove as many limits as possible from a potential response.

Quantitative and qualitative methods thus complement each other *precisely because they are based on opposing principles*. They are indeed almost direct inversions of one another, and each is especially good at what the other cannot do. Unfortunately, there is often a strong association between more quantitative methods and objectivity or detachment—replicating the view that the experimental method is the most reliable and "scientific" means of producing knowledge. This view is doubly unfortunate, both because it obscures all the "nonobjective" components of the experimental method that make it so exciting and emotional for scientists, and also because it assumes that "good science" cannot, by definition, include the human sciences within "science proper."

Working with scientists and clinicians reproduces this divide very starkly, as the anthropologist in a biomedical context will be engaged in a form of research almost antithetical to that which defines the clinical context—in which high quality medical-scientific knowledge and "reliable" data are at a premium. The following extract from Sarah Franklin's first visit to the PGD clinic in London makes this very apparent.

> Everyone is meeting for the monthly lunchtime PGD presentation in a large room with a round table that fills with about 20 people. Peter, the head of the clinic, introduces me and says he has been to my web page and does not understand a thing on it, but that the study has been peer reviewed, "so it must be kosher." I feel a bit apologetic about the very open-ended, interpretive, and "vague" nature of "an ethnography." Afterwards someone mentions that Peter has a sign on his door that says "In God We Trust. Everyone else bring your own data."

[4] There is a distinction between "closed" and "open" questionnaires, the latter referring to questions that allow participants to elaborate rather than select among preset options. Often questionnaires mix these types of approaches.

In this extract, recording a key point in the study on the occasion of the principal investigator's first actual visit to the London clinic, there are clearly several things going on simultaneously, some of which are somewhat contradictory. For example, the distancing emphasis on the incomprehensibility of Franklin's personal web page is mainly being used as a friendly joke, both to welcome her and to officially endorse her project. She is thus being welcomed to the team, and incorporated, through a form of gentle mockery that foregrounds her "strangeness" in order to integrate her. Reassuringly, this gesture also says "we may not always understand each other but we will work together out of mutual professional respect"—an excellent and very sound basis for collaboration (which indeed it proved to be).

The ability of field notes to capture the ambiguity and depth of moments such as these is what makes them valuable as a recording device for several reasons. For the fieldworker, they enable changes over time to be traced—such as the beginning of a transition from being an outsider to an insider depicted here. They also convey mood, atmosphere, and subtle social dynamics captured in dialogue, for example through humor, irony, or tone. Perhaps above all they are often engaging to read, all the more so because their immediacy conveys the "feel" of an encounter.

Fieldwork in the Clinics

It is a cliché that gaining access to "elite" field sites such as hospital clinics or scientific laboratories requires diligence and careful preparation. This process is also extremely time-consuming, and when initiated as part of an application for funded research, it must be completed far in advance of the project's potential start date. For this study, the clinicians who became coapplicants were contacted in London and in Leeds in 1999, more than two years before the project began.

Since a researcher's chances of acquiring funding are enhanced by evidence of support from cooperating clinics, it is necessary to gain clinicians' agreement to participate in a project that does not yet exist—often with a researcher they have never met or heard of, who is working in a completely different field—many months, or even years, in advance. This odd situation requires considerable good faith from both sides.

Once competitive funding has been successfully negotiated, there are numerous formal approval procedures that must take place to enable researchers to gain access to field sites. In particular, there are an increasing number of ethical approval protocols that must be satisfied. For this study, ethical approval was needed from the hospital ethics committees in both London and Leeds, and to satisfy their requirements

Patient Information Sheets about the study had to be drafted, and later redrafted, for both sites (see appendix). It was also necessary to clarify for ethics committees what consent procedures would be used, both for observations and interviews of patient, and for fieldwork in clinics involving conversations with clinicians, scientists, and health professionals. It was an additional requirement for this study that the investigators be added to the Human Fertilisation and Embryology licenses for both clinics.

Clinics are busy places where time is always at a premium, and space is often crowded and overused. To enter the space of the clinic as a participant-observer can be an intimidating and confusing process, often creating a sense of awkwardness and self-consciousness. In large teaching hospitals such as the Leeds General Infirmary or Guy's and St. Thomas', where fieldwork for this study was undertaken, it is routine for observers to be present at consultations and clinical procedures, although patients are always asked in advance if they are willing for observers to be present. Hence, at one level, a social scientist can easily "blend in" and simply become yet another visiting observer.

This sense of "joining in" is enhanced both by the strict separation between patients and clinicians that defines clinical encounters, or roles, and by the strong sense of teamwork that characterizes much clinical practice. Hence, entering a room, notebook in hand, alongside a clinician to observe a consultation inevitably feels very marked by the dividing line between patients and clinicians, and by the sense of being clearly positioned as "part of the clinical team." At one level, this makes things easier, as one's "ambiguous" or "outsider" status is erased by the sheer either-or of the setting: either you are (or are presumed to be) a medical professional, or else you are a patient.

It is equally possible to be positioned on the other side of this line, when, for example, sitting in the waiting room with other patients before being escorted back into the clinical office by one of the staff. Such situations, and movement between them, emphasize the ambiguous identity of being neither clinician nor patient.

The awkwardness of this ambiguous status is strongest at the outset of fieldwork, when, as for a new member of staff or a new patient, the setting is unfamiliar and somewhat intimidating. Over time, learning the names of staff and simply spending enough time in the clinic to become more familiar to the people who work there strengthens a sense of belonging and builds confidence. Adapting to the pace and the style of inhabiting clinical space, where everyone is rushing about performing their duties, can feel like being uncomfortable at a party and wishing one had a "task." Finding a place to sit, bringing things to read, and purposefully recording details in one's notebook can create a sense that you too have a job to do and are busy fulfilling it. The alternative, of slightly embarrassed and self-conscious inactivity, is most unpleasant.

However, while it is possible to feel part of a team, and to gain confidence about inhabiting clinical space less awkwardly, the clinical ethnographer is always a guest. Inevitably, a sense of gratitude and indebtedness thus accompanies the privilege of being "invited in." A sense of dependency and a desire "to fit in well" is inescapable, affecting not only one's demeanor and behavior but also raising concern about one's appearance and "making a good impression."

All these features of ethnographic fieldwork in clinical settings contribute to a form of "double consciousness," which is common to much anthropological fieldwork but has some aspects that are particular to an institutional or professional setting. A common experience of ethnographers in general is the sense of being an "impostor," a "spy," or "in disguise." This feeling of "ethnography as masquerade" is in some senses inevitable for all the reasons described above, which require a self-presentation that is sensitive to, and successfully adaptive within, an unfamiliar setting to which the ethnographer, by definition, does not belong. This can take the form of feeling that one is complicit in an uncomfortable deception when, for example, patients ask questions that clearly presume one is a medical professional.

The feeling of being entirely dependent on one's hosts for the privilege of being granted access to a highly restrictive space, and the desire to "fit in" as much as possible, inevitably affect the ethnographer in ways that must be considered carefully. Such a position is clearly subordinate, and ethnography can be both highly embarrassing and somewhat infantilizing. The expertise that "got you in" is all but invisible once you are there and is unlikely to be recognized more than minimally during the course of actual fieldwork, where misrecognition is much more common. More often, one is dealing with the absence of a "proper" role, misidentification as something else (doctor, nurse, social worker, psychologist, patient), and a sense of "faking it" to fit in—while even what is being imitated, exactly, is not clear, other than that one is legitimately occupying the space with appropriate authority to do so.

All these factors contribute to the special character of ethnographic exhaustion, which can be difficult to explain, even to other social scientists. What is tiring about the intense immersion of ethnography, after all, is what makes it work. By operating within and across several different levels of interaction, while being deprived of many of the aspects of one's identity normally used to negotiate them, a specific form of attention becomes possible in which proximity is combined with distance. It is important to keep track of the dissonances this form of attention makes visible, as they are where ethnographic learning takes place. Field notes are an essential means of tracking the changes in one's own perceptions of events, which, in turn, is how the ethnographer uses him- or herself as an "instrument" to gather data.

Many aspects of integrating oneself into the clinic involve adaptations that comprise important learning contexts for the participant-observer. Time and space both have very particular meanings in clinics, for example. Clinic time is extremely pressured, and clinicians often struggle to keep to busy schedules. The time frame of clinical attention is consequently very intensely "in the moment" as it moves rapidly from one context to the next. It is the chart in front of the clinician, the patient being examined, the colleague with an urgent message, or a pager beeping that sets the almost impossible pace of a clinic. Time in an IVF or PGD unit is also very tightly scheduled for both the clinician and the patient, who are working to very precise protocols of hormone injections to stimulate ovulation. An urgency of time is especially evident once eggs have been fertilized and must be biopsied for PGD, in order to be genetically tested. For clinicians, the time of the clinic is often "impossible," as their overly packed schedules cannot be met and events that are out of their control, such as procedures that take longer than expected or records that are mislaid, cause unavoidable delays. The urgent time of the clinic, in which everything is "fast," is very different for patients, for whom the waits between appointments, and in particular the wait for results between procedures, can feel interminably slow. These aspects of being inside or outside the clinic are not only important features of the professional culture of biomedicine, or the experience of being a patient; they also reveal important sociological and emotional forces at work, which literally come to be embodied by the ethnographer (Emerson et al. 2001; Bloor 2001; Murphy and Dingwall 2001).

The Uses of Field Notes

Given the intensity of ethnographic fieldwork, it is not always possible or appropriate to take notes, and they often must be compiled from memory afterward. This is one of the most difficult aspects of fieldwork, as one is often so exhausted afterward it is impossible to concentrate. Fortunately, ethnographic data is often very rich, and a little bit goes a long way. It is easy to get too much data, and over time one learns what are the key observations it is useful to record.

Field notes are essential for a wide range of reasons, and are put to many different kinds of uses, not all of which are visible in the "finished product"—the ethnography—because the primary use of field notes is to transform observations and experiences into written form. This both preserves the actuality of fieldwork and objectifies it by externalizing it, and literally turning one's experiences into an object (a text).

Field notes record everything from the emotional ups and downs of fieldwork to questions, problems, or uncertainties. They can be used to record events, people, places, and procedures. Often they are used to record a setting, or a sense of location, which may be useful in terms of remembering both the "feel" of a particular place and details about a particular visit. Later, extracts from field notes may help readers share a sense of the environment in which fieldwork took place. For example, on first visiting the St. Thomas' clinic, Sarah Franklin described the waiting room as follows:

> The clinic is on the 7th floor of the North Wing of the Hospital, which is accessed from the ground level via a crowded lift. The waiting room is large and appears previously to have been a ward, as it is still possible to see where six different beds would have been positioned from the curtain rails still in place along the ceiling. Typically for a large NHS hospital, there are an assortment of well-worn chairs and small coffee tables covered with clinic brochures, many of which are outdated. Scattered amongst them, in contrast, are multiple copies of the current issue of a glossy drug company–sponsored magazine *Pathways to Pregnancy*, which turns out to be very informative about the British IVF scene and features an article by two of the consultants in Leeds. On the wall are two large faded signs with the capitalised instructions: "PLEASE LET THE OFFICE KNOW YOU ARE HERE" and "IF YOU ARE HERE FOR AN EMBRYO TRANSFER YOU MUST HAVE A FULL BLADDER." Next to the water cooler, a 1996 HFEA Bulletin is tacked up reassuring patients that the suggestion in a BBC programme, "Making Babies," that embryo freezing might have deleterious consequences for offspring is not backed up by any research of which the Authority is aware.

As well as depicting a place or setting, field notes can help capture an atmosphere, thus giving a further sense of place to the proceedings. Because much of the "action" in a busy clinic is a mixture of formal, procedural, scheduled activities with an informal subtext of more chaotic "unscripted" incidents, field notes are an essential means of making sense of the often complex logics of practice that define spaces that are both highly ordered and ordinarily "messy." They are also helpful for data analysis in terms of re-creating a scene or interaction, by giving it a context and a literal background.

Field notes are thus both aide-mémoire and a means of processing field "experiences" that help clarify them. They preserve nuances that are hard to capture, and which illustrate the culture or way of life of a particular setting such as a busy clinic. They re-create the "real time" of social practice and exchange and a sense of "being there." In the same way, they personalize encounters and give a sense of interaction. By restaging the before, during, and after of ethnographic work through re-created incidents, they allow a sense of the process of research to be conveyed through the accumulation of experience and detail. They allow for the more colloquial speech of everyday encounters to play a part in what is often a more formal language of academic writing. This adds a dimension to academic writing that is distinctive because it allows for the writer to be differently present in the text.

Although field notes are useful in all these ways—because they are immediate, specific, and convey a sense of feeling—these attributes also make their use problematic, as they can be overly revealing, overly personal, or sentimental. There are also practical problems. For example, the necessity to anonymize patients meant that no descriptive details could be provided about any of the PGD couples interviewed for this study. Similarly, many events observed in clinic can be recorded for the purpose of personal reflection and record keeping but cannot be used in print, as they concern confidential interactions.[5] As Paul Atkinson observes, there are many variations in tone in field notes—some ethnographers adopt a confessional voice, others a more heroic attitude. These choices depend somewhat on the kind of setting being described, and how strictly information is controlled within it.

Interviewing Patients

In any study of a topic as intimate and personal as PGD, the practice of collecting data in person from patients raises innumerable ethical, emotional, political, and practical issues. While many of these can be taken into account and prepared for, there is no fail-safe method even for how to contact patients to begin with, and so constant vigilance, double- and triple-checking of procedures, and consultation with clinical staff are essential.

This study was facilitated by the PGD coordinators at both London and Leeds, although patients were contacted only through the London clinic for interviews, owing to unexpected delays in beginning PGD treatment at Leeds during the course of this study. Letters were sent out by the London PGD coordinator, Jenny Caller, with a description of the study to all PGD patients coming through the clinic. A reply form indicating a

---

[5] The "scholarly price" of using field notes in ethnographics of professional settings is discussed by Rabinow (1999, 6).

willingness to be interviewed, and supplying contact information, was provided along with the Patient Information Sheet and a stamped envelope addressed to the research team. Contact details were also provided on the Patient Information Sheet as well as the project website, in case anyone had questions or wanted to see the research proposal.

Given the very small numbers of PGD patients being treated in all of Britain on an annual basis, it was clear from the outset it would be likely to take at least a year to interview even ten couples. In the end, twenty-three PGD patients were interviewed for this study between January 2001 and August 2002, all of them in their homes, and all in England.[6] Interviews began with a description of the study, followed by an explanation of the key points on the Patient Information Sheet, for example anonymity and the option to receive a transcript of the interview. All interviewees were asked to sign a Consent Form agreeing that the interview could be tape-recorded and transcribed and extracts could be used in publication.

The interviews themselves were open-ended and semistructured, meaning that a roughly chronological set of questions was used as a guide through the experience of undertaking PGD. (How did you find out about PGD? What was your first consultation like? Were you aware of what was involved in IVF? Etc.) Depending on the circumstances, these interviews lasted between one and two hours. Although often very emotional, patients often commented that the interviews were helpful in enabling them to reflect on their treatment and to talk through aspects of the personal experience of PGD in more depth. Being an interviewer in such a situation is nonetheless a very demanding role. While the intimacy of the content of the interview creates a strong bond among the participants, the larger context is an academic research project, with which patient-interviewees have only a brief encounter. Similarly, being an interviewer puts one in a complicated emotional position, as one must share very intimate, and often upsetting, conversations with patients whose lives one is entering for only the briefest of glimpses. This asymmetry is one of several challenging features of interviews, which by their very nature transgress conventional social boundaries between the public and the private, mixing the domain of personal experience with that of professional activity. Practically they also present challenges, in terms, for example, of the timing and direction of questions. The goal of maintaining an open-ended space of conversation, in which the interviewees are free to follow their own thoughts, has also to be managed through questions that are phrased and timed to provide

---

[6] Although the difficulty of securing PGD treatment in the United Kingdom means that the St. Thomas' clinic treats patients from as far away as Scotland, Ireland, and Northern Ireland, it was not feasible to arrange interviews with patients traveling from such great distances.

continuity. Thus, the interviewer's task is to prompt as much as to question, often, for example, simply by repeating what the interviewee has just said.

Transcription and Citation

While the process of recording field notes is often done by hand using notebook and pen, the transcription of interview tapes is a more elaborate process involving a transcription machine and a computer. By using foot pedals, the speed of the tape can be controlled, leaving both hands free to transcribe the tape-recorded speech word for word. Although laborious, transcription is fascinating, as, with the benefit of headphones obscuring all other sounds, one is reentered into the space of the interview, conversation, or event—in some ways even more intensely for *not* being physically present. Not being physically present allows one to hear everything undistracted by the mundane but often preoccupying details of conducting an interview. This process is intensified by the concentration necessary to distinguish each word from another and transcribe them. Inevitably one not only hears but comprehends much more than one did initially. Frequently this is accompanied by sensations of disappointment and remorse, as one hears oneself interrupting the interviewee just before he or she sounded, in retrospect, to be on the verge of saying something particularly interesting.

In addition to the literal content of recorded speech, which is intensified by transcription, there is a formal content evident in the pace, or flow, of speech—such as in hesitations, accelerations, and changes in tone or volume. This aspect of speech is unfortunately diminished by transcription, which reduces voices to words. An advantage of using a relatively small number of transcripts, meaning fewer than twenty, and ideally closer to ten, is that it is possible to keep a sense of the flow of the conversation, the speaker, and the setting in mind, as well as its literal content. The possibility of retaining even a weak connection between speakers, voices, and texts is highest when the transcription pile is smallest. Likewise, when more than twenty separate transcriptions have been collected, some of which can be up to sixty pages long, it becomes increasingly difficult to make full use of the data. Here too is one of the ways in which qualitative data differ from quantitative: more is not necessarily better. The ability to see the data set "as a whole," and to reflect back over whether one group of interviewees was really saying something quite different from another and what these differences were, is diminished in direct proportion to the quantity of data.

What is remarkable about transcribed extracts of interviews and

TABLE 2.1
Text Etiquette

interview	recording	transcript	extract
speech	words	text	text-ette

At every stage, the translation of an interview into text is a process of reduction.

conversations is how much people enjoy reading them. Despite the fact that an extract from a transcript is several degrees of separation removed from actual speech in real time (see table 2.1), it retains a great deal of its vitality nonetheless when reproduced.

Once a tape has been transcribed, the transcript must be "cleaned" and checked for errors, of which there are inevitably many. It is surprisingly difficult to produce high-quality recordings using either analog or digital recorders. There are advantages to using both—for example two analog recorders with battery-amplified microphones attached to interviewees' clothing whenever possible, and a digital recorder as backup (it has only recently become possible to transcribe directly from digital recordings using specialized equipment and software; previously they had to be rerecorded onto analog tapes to be transcribable).

While it is desirable to retain as much speech-as-it-is-spoken, extracts for publication in this book have been "cleaned": a higher than usual percentage of the ums and ahs were removed, along with some of the more fragmented sections between sentences. "Cleaning" also involves some grammatical corrections, in particular where the speaker could be cast into an unflattering light or perceived in a stereotyped manner. Such decisions are far from straightforward and are perhaps best described as textual etiquette.

The textual etiquette used for this study attempted to present the speakers faithfully but also courteously and respectfully. To arrive at a suitable compromise involved experimenting with different methods of referring to participants in the study and "adjusting" the anonymity "volume." As noted in the acknowledgments, it was decided at the outset to use proper names as much as possible and to further identify speakers for readers using footnotes wherever possible. As also noted earlier, one reservation was the division this created between the re-named people who spoke about their personal experiences of PGD and the properly named people who had a professional relationship to "the world of PGD." This contrast exacerbated a related difficulty of whether or not to refer to people who spoke about their experiences of undergoing PGD as "patients," as "couples," or merely by their pseudonyms. In the end, all three strategies are used in this book, each of which has disadvantages: "patients" can be seen as a reductive and pejorative label (although possibly less so in Britain than in the United States); "couples" is slightly euphemistic and inaccurate in phrases such as "couples expressed the view" or "couples who became pregnant," as if two people were each part of the couple in the same way; and pseudonyms, with their emphasis on secrecy and concealment, could be seen to be disrespectful by their very nature.

Such are the unavoidable difficulties of ethnography, which, as noted earlier, is far from a democratic or egalitarian method.

Following PGD Around

Whereas clinical ethnography and interviews provide a portrait of PGD that is up close and personal, "experience near," and often dominated by technical details, the view from the media or public debate is far more remote. High-profile cases of PGD for couples seeking to provide a "savior sibling" for an existing child with a chronic degenerative disease, as in the Hashmi case, dominated media headlines and radio chat shows during the period of this study. Meanwhile, the connection between PGD, embryo research, and stem cells occupied the minds of policy makers and parliamentarians. As described in the previous chapter, the image of the "designer baby" was as varied as it was ubiquitous during this period, indexing both social anxiety toward PGD and hope for its potential medical benefits.

The archive of PGD representations assembled from these sources—compiled by "following PGD around" to conferences, parliamentary committees, and conferences, as well as collecting examples of media coverage—illustrated a wide range of conflicts over its legitimacy and future use that provided a crucial background to the interview and participant-observation data. These versions of what might be called "parliamentary PGD" or "media PGD" or simply "public PGD" were often quite different from the actual uncertainties and negotiations encountered in the context of clinical practice, thus establishing one of the major gaps this book is designed to explore—between PGD as a desirable form of individual clinical treatment and PGD as a symbol of reprogenetic anxiety more generally.

This emphasis on PGD in public derives from the view, well established within cultural studies, that it is not possible to separate private, individual, or expert opinions on controversial topics from the effects of more widespread public conventions of representation. In this view, PGD is both a medical technology and a discursive technology, in that it is both a specific technique and what, in cultural studies terms, would be described as a contested domain. As both a medical procedure and a cultural symbol, PGD acquires an overdetermined significance that is derivative of its multifaceted meanings.

An Ethnography of PGD

To summarize, then, traditional ethnographic approaches were adapted for this study of PGD both by using conventional methods of participant-observation to conduct fieldwork in the clinic and by supplementing this with interviews. In addition, cultural-studies approaches of examining popular media representations of PGD, and the depiction of "PGD in public," have been used to provide a backdrop of public debate that links back to some of the themes raised in the previous chapter about the distinctive role of PGD in shaping legislation concerning, and governance of, human fertilization and embryology. The overall strategy, then, was to collect understandings and representations of PGD from a range of contexts, and to use these as comparative "frames" for each other (see J Edwards et al. 1993, 1998).

Many of the themes that organize the findings of this study, including the emotionality of PGD, its complicated relationship to hope for scientific progress, the labor involved in its provision, and the value of uncertainty during treatment, could have been identified only through highly qualitative and open-ended methods. The considerable contrast between the views of PGD expressed in the range of different contexts is also a finding that belongs to the specific form of apprehension associated with a sociological perspective. Indeed these results are in many ways very typical and obvious in this respect.

While the advantages of such an open-ended study are that it can be radically exploratory and almost indiscriminate in terms of what counts as "data," the disadvantages lie in the ability to test a particular hypothesis in a more systematic manner. At this point in time, when the social science of biomedicine and genomics is only beginning to build up a methodological "tool kit" and a sufficiently broad range of studies to develop longitudinal and comparative analysis, both qualitative and quantitative forms of research are required. Studies such as this one will identify unanticipated areas of potentially valuable research variables, which more focused large-scale studies can investigate more systematically.

Chapter 3
Getting to PGD

Preimplantation Genetic Diagnosis (PGD) is an evolving technique that provides a practical alternative to prenatal diagnosis and termination of pregnancy for couples who are at substantial risk of transmitting a serious genetic disorder to their offspring. Samples for genetic testing are obtained from oocytes or cleaving embryos after in vitro fertilization. Only embryos that are shown to be free of the genetic disorders are made available for replacement in the uterus, in the hope of establishing a pregnancy. PGD has provided unique insights into aspects of reproductive genetics and early human development, but has also raised important new ethical issues about assisted reproduction.
—Peter Braude, Susan Pickering, Frances Flinter, and Caroline Mackie Ogilvie, "Preimplantation Genetic Diagnosis"

In the same way there is no single definition of PGD, so too there are several different ways of "getting to PGD"—for scientists, clinicians, genetic counselors, patients, and researchers. Each of these involves a different path of approach, and consequently a different sense of arrival. Paths to PGD clearly depend on where one is located to begin with, and the first section of this chapter presents a brief account of the picture that has emerged after PGD's first decade of clinical practice, in which access to it was both widened and heavily restricted. This overview also helps to situate the clinics involved in this study in relation to their wider professional associations, and in relation to some of the key issues that define the PGD field internationally.

The first "arrival" described in this section is thus of an emergent subfield of reproductive biomedicine, which had, by the year 2000, most definitely become established on the world scene as a prominent and influential context of technical innovation, scientific research, and clinical practice. It had also acquired a particular profile within the United Kingdom—in terms of both its technical practice and its rate of expansion.[1]

This chapter also examines how patients "get to PGD" in terms of how have initially encountered it, and the decisions, hopes, and chance

---

[1] A more complete technical description of PGD is provided in chapter 4.

events that have brought them to the clinic for their first consultation. Here, the "arrival" at PGD is narrated in terms of how people have progressed from not knowing anything at all about PGD to walking through the front door of the clinic for their first appointment. From the clinical side, the attitudes of clinicians, nurses, genetic counselors, and PGD coordinators toward patient perceptions of PGD, and the kind of choice or option it offers, are explored.[2] These perspectives portray both the difficult challenges involved in pursuing PGD and the ongoing expansion of these challenges, with the emergence of options such as HLA typing through PGD to create a "savior sibling." The next two chapters investigate how this difficult and challenging world of PGD is inhabited by its "insiders," and in particular how they describe and manage the difficulties they face.

The Birth of the PGD Clinic

Unlike IVF, which has expanded at a rapid rate ever since its initial success in 1978, the evolution of PGD has been more gradual and restrained. There are several reasons for this, which are explored in some detail below. It is, however, somewhat difficult to tell the "story of PGD," both because it is complex and multifaceted, and because there are many different versions. Indeed another book could easily be written on the worldwide development of PGD, and both historical and comparative studies would be of significant value in this field. Even in the "official" medical literature there are many different accounts of how PGD emerged, and interviews with key figures involved in the development of PGD only emphasized this complexity. While it is necessary to qualify the following account of how PGD has "come of age" by noting it is necessarily a point of view, this chapter would be incomplete without at least a brief account of some of the distinguishing characteristics of PGD as a form of clinical practice, since these are highly relevant both to the experiences of patients undergoing PGD and to its regulation.

Following its first successful performance in 1990, by Alan Handyside and Robert Winston at London's Hammersmith hospital for a case of Duchenne muscular dystrophy sexing, PGD expanded steadily, although gradually, as it became less associated with experimental medicine or research and more widely accepted as clinical practice. By the year 2000, its tenth anniversary of sorts, PGD had, like IVF before it, become a major

[2] This chapter does not explore the question of how clinical professionals "arrived" at PGD, which would have been an additional and very useful perspective. However, the usual constraints of time and space precluded this option, which is the subject of ongoing research.

international field of reproductive biomedicine. However PGD developed much more slowly than IVF. By 1997, forty PGD centers were established in seventeen countries, and approximately 1,200 cycles of PGD had been performed, resulting in 231 pregnancies and 166 healthy births (Verlinsky and Kuliev 1998, 215). In their *Report* published during the first year of this study, 2001, the International PGD Working Group reported over 2,000 clinical cycles, resulting in 531 pregnancies and 395 healthy children, of whom 289 were preselected to avoid chromosomal abnormalities and 106 to avoid single-gene disorders (International Working Group 2001).[3] According to data published by the European Society for Human Reproduction and Embryology (ESHRE) PGD Consortium in 2002, 1,197 cycles of PGD had resulted in a 22.4% pregnancy rate (ESHRE PGD Consortium 2002).

However, while the expansion of PGD can be assigned a healthy growth rate of over 100% per annum through such accountings, with corresponding annual increases in the range of different treatments provided, these "global" figures mask important features of the overall PGD "picture" in terms of clinical practice. For example, more than half of the world's PGD cycles are performed in one center, the Reproductive Genetics Institute (RGI) at the Illinois Masonic Medical Center in Chicago, headed by Yuri Verlinsky. Moreover, the vast majority of cycles in Verlinsky's clinic are performed for what is known as age-related aneuploidy, referring to the rising rate of chromosomal abnormality in the egg cells of women over thirty-five ("of advanced maternal age"), which typically compromises the IVF success rate.[4]

This use of PGD in conjunction with IVF—in essence as a fertility enhancement measure—is much less commonly practiced in Europe. Indeed, at the leading European PGD clinics in London (Hammersmith, University College London, and Guy's and St. Thomas') and Brussels, the use of PGD to screen for aneuploidy is minimal, and is referred to as "PGS" (Preimplantation Genetic Screening) or "PGD-AS" (PGD for aneuploidy screening) to distinguish it from PGD involving couples who

[3] The first International Symposium on Preimplantation Genetics was held in Chicago in 1990, at which an International Working Group on Preimplantation Genetics was formed, chaired by Yuri Verlinsky. This group subsequently reconvened at the annual meetings of the European Society for Human Reproduction and Embryology in 1993 and 1995 (and in 1994 and 1996 in conjunction with other international conferences). A second International Symposium on Preimplantation Genetics was held in Chicago in 1997, the same year in which a PGD Consortium was established within the European Society of Human Reproduction and Embryology (ESHRE).

[4] According to the February 2004 figures on the RGI website, over 100 cycles involving genetic disease have resulted in the birth of two dozen children, while screening for common aneuploidies has been offered to 600 IVF couples, resulting in 131 births of healthy children (www.reproductivegenetics.com).

are usually normally fertile but have a recurrent risk of transmitting severe genetic disease.[5]

This division in part reflects an effort to associate PGD more closely with genetic medicine than with infertility treatment, and with genetic screening rather than assisted conception. As Peter Braude is at pains to make clear to his audiences at Patient Information Evenings (PIEs) at Guy's and St. Thomas' Hospital, "PGD is a means of treating genetic disease, not a form of assisted reproduction."[6] As is evident from chapter 1, the history of debate in Britain might be interpreted to suggest that IVF and infertility "alone" are less medically or scientifically defensible than the possibility of avoiding lethal childhood genetic disease as grounds for justifying the need for embryo research to a skeptical and divided public. It could also be said that since IVF is the "weakest link" of PGD treatment, it makes sense for its role in PGD to be de-emphasized. Likewise, in terms of convincing health authorities to fund PGD, its associations with IVF could be seen as detrimental in that they make PGD look like a form of fertility treatment, rather than genetic disease prevention.

It is for similar reasons that the rapid expansion in the use of PGD for aneuploidy screening is a cause for concern among PGD specialists in Britain. If, as has been confirmed in part by the increasing clinical use of PGD, a significant percentage of IVF cycles are predestined to fail even before the "best-looking" embryos are transferred, because hidden in their nuclei are fatal chromosomal errors, then the use of PGD to detect chromosomal anomalies (PGD-AS)[7] would clearly increase the chances of successful pregnancy.[8] Moreover, since many women attempting IVF are over thirty-five—and an increasing number are closer to forty-five— there is a potentially very large and profitable customer base for PGS. However, rapid growth of PGD-AS could have damaging consequences for PGD, by more strongly associating PGD not only with IVF *but specifically with the high rates IVF failure that make this association problematic to begin with*. Because it is potentially a very commercially profitable "niche" in the market for reproductive services, PGD for

[5] As noted earlier, preimplantation diagnosis (PID) first became preimplantation genetic diagnosis (PGD) and then evolved into preimplantation genetic screening (PGS), now officially further designated by the European Society for Human Reproduction and Embryology PGD Consortium as PGD for aneuploidy screening (PGD-AS).

[6] In contrast, for example, Robert Winston writes: "Pre-implantation genetic diagnosis— that is genetic diagnosis in an embryo before implantation—is the first application of IVF for a condition other than infertility" (1999, 83).

[7] Somewhat confusingly, although the ESHRE PGD Consortium has designated PGD for aneuploidy screening PGD-AS, the HFEA still use the acronym PGS.

[8] In a 1999 study across three different centers, PGD-AS was shown to decrease the rate of IVF failure following embryo transfer from 25.7% to 14.3%, with a corresponding increase in live healthy births from 10.5% to 16.1% (Munne et al. 1999).

aneuploidy screening (PGD-AS) could also be seen to have negative associations with the increasing incidence of "special treatment" for those who can pay privately—a controversial trend that is seen to have created a "two-tier" system within the National Health Service.

In the United States, where there is no public health care system, no licensing authority, and less regulation, and where commercialization is less controversial, there is less of a conflict between the use of PGD to improve IVF treatment and its use to prevent genetic disease. In Britain, however, a key component of the Warnock strategy for minimizing public "discomfort" with new reproductive and genetic technologies has been to curtail, if not ban outright (as in the case of commercial surrogacy), their association with profiteering and its potentially exploitative consequences. These national differences form an important part of the background to debate over PGD, much as they do in other areas, such as IVF, stem cell research, cloning, and surrogacy.[9]

The developing PGD "picture" since 1990, then, can be viewed in terms of a number of important contrasts. Although PGD began as a form of experimental medicine, its repeated success (and the lack of any indication of damage to offspring from embryo biopsy) has established it as a medically accepted alternative to prenatal screening, though not as a routine procedure. While there are an increasing number of PGD clinics worldwide, there is a sharp divide between the half dozen most successful centers, where the majority of PGD cycles are performed, and the much larger number of smaller centers, where treatment cycles are more infrequent. Both large and small clinics operate in distinctive national climates affecting their practice. At the world's largest clinic in Chicago, most PGD is undertaken to improve IVF success rates.[10] Similarly, in Australia, where an increasing number of PGD clinics have been established, a significant portion of PGD cycles are performed for sex selection—a pattern also seen in India, China, Israel, and the United States.

Britain's place in the wider global context of PGD can thus be described as medically and scientifically progressive, but cautious in relation

[9] Owing to the tiny number of cycles undertaken annually in the United Kingdom, it is difficult to generalize about the funding situation, save to say that in some cases a specified number of PGD cycles will be paid for by the Local Health Authority, whereas in other cases it refuses to offer any financial assistance. As a result, between a half and two-thirds of PGD cycles in the United Kingdom are privately funded, at a cost of as much as £5,000 per cycle. In cases where the Local Health Authority has refused to give couples support, the clinic may do what it can to minimize costs, such as passing on to patients the lower costs of pharmaceuticals purchased by hospitals.

[10] The RGI relies largely on genetic and chromosomal diagnosis of the first and second polar bodies. This technique can be used to detect only maternally derived disorders. A different method, biopsy of the embryo at the eight- to twelve-cell stage, is more commonly preferred by many PGD centres (Braude et al. 2002, 944).

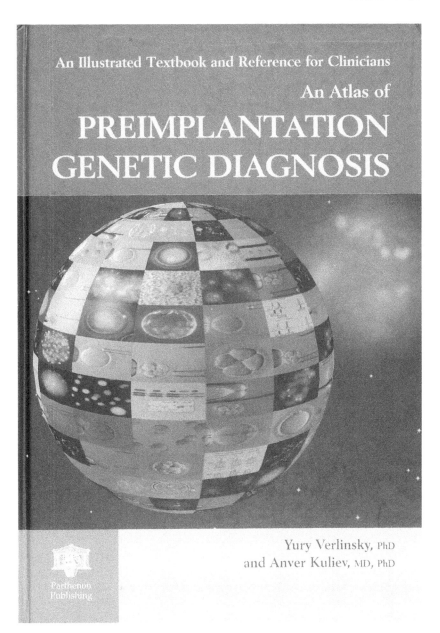

FIGURE 3.1. This image of a planet that resembles Earth, orbiting in a galaxy of fluorescent stars and composed of diagnostic imagery from PGD, recalls the photographs of the "blue planet" taken from outer space, here transformed into an image of "inner life."

99

to uses of PGD that could be seen as too commercial—including the "boutique" choice of sex preselection and the high-priced niche market in aneuploidy detection. Although Britain is not the largest provider of PGD in global terms, its scientific contributions to the technique have continued to be distinguished since PGD's first successful implementation in London in 1990. This is in part because the medical-scientific community in Britain has sought to protect the image of PGD by minimizing its association with fertility enhancement and emphasizing its "original" role as a branch of clinical genetics.

"Slow Progress"

As is evident from the early accounts of PGD described in chapter 1, its "prospects" were viewed with guarded hopes in the 1980s even by leading advocates of its development, such as Anne McLaren. Similarly in the early 1990s, Robert Edwards described the situation as one of "caution and optimism." As he wrote, "there are well based grounds for cautious optimism" (1993, 19), since the technique is "likely to appeal to many parents" and "elegant methods are now available in experimental biology and molecular biology" (19). However, "there is an equal need for caution," he added (19). Not only do "embryos themselves signal some cautionary tales," such as their capacity to express themselves mono- or trisomically, or with "even more extreme chromosomal anomalies which could lead to errors in interpreting diagnoses" (19), but also their unreliable tendencies toward mosaicism, or duplicity, mean that some parts of the embryo may "send misleading information" about its overall constitution.[11] As Edwards pointed out, "DNA technology is not perfect either" (19), especially when applied to minute quantities of material removed from a single cell. Errors can occur in amplification, through contamination, or both. Highly accurate, reliable, precise, and verifiable assays, which are the bottom-line of PGD, can be made maximally error-resistant *but never fully error-proof*. The technical challenges of PGD are thus both enormous and enormously varied, but as Edwards was keen to emphasize, these are matched by the ongoing challenge to comprehend the embryo's own complex internal expressivity—much of which has been further revealed through PGD.[12]

[11] Robert Winston makes a similar point in his 1999 description of PGD, from his popular guide to assisted reproductive technologies, *The IVF Revolution*, in which he begins his section "Biopsy of Unrepresentative Cells" with the statement: "Human embryos are frequently defective in all manner of ways" (1999, 95).

[12] Some of the aspects of the embryo's own "expressivity" noted by Edwards are "genetic imprinting" and "genomic reprogramming." He also mentions the problems of allelic

This tendency of PGD experts to emphasize the error-prone nature of the technique is echoed in Robert Winston's lengthy description of PGD in *The IVF Revolution* (1999), a book that offers a frank and uncompromising view of assisted reproduction as both "increasingly sophisticated" but "still quite crude" (1999, vii). Among the subheadings under "Problems with Preimplantation Genetic Diagnosis" are "Contamination," "Failure of Amplification," "Biopsy of Unrepresentative Cells," and "Damage Caused by Embryo Biopsy" (1999, 94–95). These sections are followed by yet another list of problems that result from the "desperation" that leads many couples considering PGD, in Winston's experience, to "put themselves at ill-considered risk." Among the considerations that must be thoroughly discussed during the initial consultation process, in his view, are the risks of implantation failure, miscarriage, twins or multiple births, misdiagnosis, inadvertent damage to potential offspring from embryo biopsy, stress, and expense (1999, 96–98). In contrast to the somewhat celebratory early accounts of IVF, including those of Winston himself,[13] the tone of much of the more recent PGD literature—be it directed at patients or professional audiences—is cautionary to the point of appearing almost "anti-PGD." As we shall see, however, such precautions are designed with a protective purpose in mind, and in particular the goal of narrowing PGD practice to its most precise uses in the interests of its steady, but gradual, progress. If there was a view of IVF as somewhat overprescribed (a position Winston himself expressed), the same cannot be said of PGD.

In a short communication in the journal *Prenatal Diagnosis* in 1998, Alan Handyside similarly described the clinical progress of preimplantation genetic diagnosis as "slow"—a result, he suggested, of both the difficulties of setting up a PGD unit capable of combining "expertise and resources in both reproductive medicine and molecular genetics . . . along with specialised facilities for handling single cells," and the "diversity of chromosomal and genetic defects" requiring new protocols or "work-ups" (1998, 1345).[14] This representation of PGD as "slow,"

---

dropout (ADO) and mitochondrial inheritance. Above all he emphasizes the problem of "inferred diagnoses," such as polar body analysis, which rely on the assumption that a part of the embryo can faithfully represent its genomic constitution (R. Edwards 1993, 19–20)

[13] Robert Winston's fluctuating views on IVF have ranged from initial opposition in the 1970s to a conversion of sorts in the 1980s, and he has become an increasingly vocal publicist of new reproductive and genetic technologies in the 1990s.

[14] In contrast, Eugene Pergament, a member of the leading US PGD team in Chicago, suggests in his 2001 foreword to *Preimplantation Genetic Diagnosis*, coedited by Handyside with Joyce Harper and Joy Delhanty, that "although only ten years have passed, remarkable advances in technology and in the science of preimplantation genetic diagnosis have occurred."

because it is beset by considerable obstacles to progress, suggests that its growth will be steady but limited—a view that has largely prevailed, especially in Britain, since even before it was first implemented successfully at Hammersmith, and a view that differs quite markedly from the depictions of PGD we saw earlier as "out of control" and "racing ahead."

Writing of the future of PGD in a 2002 textbook, *Assisted Reproductive Technology: Accomplishments and New Horizons*, David Cram and David de Kretser of Australia's leading PGD program, based at Monash, suggest additional problems affecting PGD, although they also express less cautious assessments of its potential expansion (see below). Despite PGD's comparatively high success rate (20% pregnancy rate with 97% accuracy of diagnosis), and despite being "the most advanced form of genetic testing [yet] developed" (195), they argue that "the procedure has not been widely embraced by patients," citing "the paucity of centres offering PGD; the lack of education of patients, genetic counsellors and clinicians; low IVF pregnancy rates per cycle (25%); and high cost" (195). Reviewing the evolution of genetic diagnostic methods for PGD, Cram and de Kester provide numerous examples of technological improvements to address challenging "expressivity" problems such as point mutations, polymorphisms, deletions, and allelic dropout. They conclude there is room for substantial improvement: "with the availability of new genetic tests, there will be many hurdles to overcome, particularly in the area of assisted reproduction and predisposition to age-specific disease," they predict (202).

This mixed report card for PGD reflects, as Handyside suggests, the inherent challenges of a technique that both combines many clinical and scientific forms of expertise and covers an enormous range of genetic and chromosomal conditions—many of which are by definition poorly understood. As Edwards adds, PGD's "slow progress" reflects the combination of technical complexity and the often unpredictable and undetectable character of genetic "expressivity," which can lead, in his words, to "cautionary tales." This list of precautions is extended by Winston in his analysis of "problems with PGD" to include the difficulty of fully informing potential patients about how many things can go wrong. And, as Cram and de Kester demonstrate, the paradox of PGD is that it is always in part "diagnostic" of the next generation of technological hurdles in the effort to prevent genetic disease.

Cram and de Kester, in contrast to many British PGD practitioners, echo the views of some of the more "extreme" commentators encountered in the introduction and chapter 1 in their prediction that PGD will usher in an automated future of whole-genome assays at birth that will detect all known genetic abnormalities and lead, eventually, to "a convenient means of having children . . . with desirable traits" (2002).[15] In a significant departure from the themes of "slow progress," "cautious optimism," or the error-prone nature of PGD, they dramatically claim:

> In this new century the stage is set for a revolution in functional genomics and human genetics that will give us a much deeper understanding of health and disease and open up new possibilities for assisted reproduction. Central to this revolution will be the gene chip . . . which will enable cost-effective, rapid and high throughput analyses of the patient's genome. . . . Eventually, total genome screening using gene chips containing all known disease alleles and trait polymorphisms will become a reality. . . . The identification of an individual's genetic blueprint at birth will provide the clinician with powerful tools to assess disease risk and implement appropriate programs and treatments to manage the patient's health and well being effectively over a lifetime (2002, 202)

This confident prediction of PGD's eventual coupling to whole-genome analysis of all newborns using an inexpensive but high-quality "gene chip" as a means of enabling physicians to "manage the patient's health and well being effectively over a lifetime" is indicative of how vast, and optimistic, its "prospects" may seem, even to practitioners well aware of its day-to-day demands and limitations. A sufficient belief in the power of genetic information, wedded to a firm faith in technological improvement, is directly predictive of PGD as the "ultimate" diagnostic, for everyone, of all known genetic conditions, at birth.

On the whole, the stance of "cautious optimism" expressed by Anne McLaren in the 1980s and by Robert Edwards, Robert Winston, and Alan Handyside in the 1990s most closely resembles the attitudes toward PGD encountered in this study, possibly indicating that this is a more typically "British" point of view.[16] Indeed, an apparent contrast

---

[15] In 2004, Ms. Chelsea Salvado, a PhD student in David Cram's Institute of Reproduction and Development in Monash, reported the successful development of a "gene chip," or microarray technology, capable of detecting one of the commonest mutations (cystic fibrosis) with 100% accuracy. However, the usual PGD difficulty of achieving sufficient diagnostic accuracy using only the contents of a single cell meant this technique needed to be conducted using ten cells from a blastocyst biopsy—casting doubt on its clinical viability. The promise of the "gene chip," while theoretically attractive in terms of disease reduction, will be particularly difficult to realize in the context of PGD, where such minute amounts of cellular material are available for genetic testing.

[16] For example, this "cautious" or "British" attitude could reflect the history of public debate over embryo research and a sense of wanting to "downplay" some of the more sensationalist accounts of "designer" babies, or it may reflect a particularly British emphasis on "modesty" as a component of scientific credibility (Shapin and Schaffer 1989). See further Franklin 1997, Haraway 1997, Squier 1994, McNeil 1987.

can be seen between a largely British view of PGD as a gradually developing technology with substantial inbuilt and external limitations to growth, and the views of those scientists and commentators for whom PGD appears to be an almost unstoppable form of increasing genetic control. As Peter Braude and his team characterize the challenge for PGD in the future in the conclusion to their 2002 review of this technique, published in *Nature Review Genetics* (the abstract for which provides the epigraph of this chapter), "The challenge will be to regulate the use of PGD technology for medical purposes and to limit or prevent its use for eugenic selection" (2002, 951).

Learning PGD

A similar set of perspectives can be found "on the ground" in terms of observing an actual PGD cycle and observing PGD being performed (or not, as in Leeds), where a similar attitude of "cautious optimism" prevails. Because PGD involves several scientific disciplines and complex techniques, its learning curves are impressively laborious—a description that also applies to the amount of expert time, attention, precision, and skill required to "do" a PGD cycle. Both of the primary technical requirements of PGD—embryo biopsy and genetic testing—are fraught with hazards and require not only considerable skill but constantly repeated practice.[17] The embryo itself is both genetically unstable and unpredictable, adding another layer of potential complications. These, and all the other difficult aspects of PGD, are why large numbers (at least six) of embryos are necessary for a good chance of success. Yet the embryos for PGD have to be produced through IVF—a technology that has a 75% failure rate that appears not to be substantially altered when it is used for fertile, instead of infertile, couples.

The basic technique of PGD, as we have seen, involves the removal of a single cell from an embryo (or in some cases two cells, or the polar body) and analysis of its minute contents through amplification and extremely sensitive diagnostic assays. While it is slightly "easier" technically to sex an embryo, rather than probe for a single-gene mutation, the number of places things can go wrong remains staggeringly high. What is also important to appreciate is how many embryos are necessary in order for PGD to have a reasonable chance of success. This means that the IVF has to work especially well to provide sufficient embryo quantity and quality for PGD—a tall order indeed.

---

[17] See Franklin 2003c for a detailed description of observing a demonstration of embryo biopsy and "learning to see" its nuclei.

Neither are the technical obstacles to PGD its only limiting conditions: as Cram and de Kretser and Winston point out, there are significant social factors that affect demand for PGD. Although PGD may become much more widely known in the future, generate more referrals, and thus be more commonly encountered by couples toward the beginning of their reproductive lives, there will still be very complex attitudes toward PGD, for all the reasons this book explores in some depth. It is consequently very difficult to predict "where PGD is headed," and, for the moment, Alan Handyside's depiction of "slow progress" most accurately describes the "broad picture" of PGD.

PGD in the Clinic

The two clinics participating in this study, coincidentally, mirrored precisely the three-steps-forward, two-steps-back PGD "stagger" or "slow progress" described by Handyside. In Leeds (where Handyside was the leader of the team), the PGD team was unable successfully to relaunch their program during the course of this study, largely for the reasons Handyside himself had earlier identified. Always on the verge of resuming treatment, and with a long waiting list of referrals, the Leeds team continually met with delays that prevented the provision of treatment. The London team stood in sharp contrast, being the busiest and most successful PGD center in Britain, and recently having completed their first one hundred PGD cycles in December 2000, shortly before this study began.

Reporting on the outcomes of the first one hundred cycles of PGD at the Guy's and St. Thomas' center, Pickering, Polidoropoulos, et al. (2003) report on the results of 397 referrals from geneticists, genetic counselors, gynecologists, and general practitioners, and on a small number of individual patients, from early 1997 to December 2000. Of these, 124 cases, or 31%, resulted from structural chromosome rearrangements, otherwise known as translocations—the most common form of genetic disorder (occurring in 0.2% of the population). Other referrals were made on the basis of autosomal dominant diseases such as Huntington's (79%) and myotonic dystrophy (21%); single-gene disorders including spinal muscular atrophy (51%), cystic fibrosis (38%), and sickle-cell disease (9%); and X-linked conditions, most notably Duchenne muscular dystrophy (23%).

In the first one hundred cycles of PGD, undertaken by sixty couples over the course of three years, the Guy's and St. Thomas' PGD team developed specific test protocols for seventeen translocations, eighteen X-linked disorders, and seven single-gene disorders—in total comprising over forty different types of treatment for as many different conditions.

105

From the first one hundred cycles, twenty-five healthy babies were born. This is an impressive success rate by PGD standards, and particularly so because of the range of different conditions being diagnosed by the Guy's and St. Thomas' team.

While for many clinical procedures a success rate of only 25% would not be acceptable, PGD, as we have seen, both poses and responds to very specific dilemmas. For a couple facing a one in four chance of having a child affected by a specific genetic disorder, the success rate of PGD is no better than the risk of passing on a serious genetic disease to their offspring. Seen from this point of view, it would appear to make more sense to attempt "spontaneous conception," to be followed by prenatal diagnosis if a pregnancy is established. However, the practicality or "logic" of such a strategy may be offset by the emotional and psychological toll from either the previous death of a child/children or repeated terminations—as in the case of Bernadette Modell's patients mentioned in chapter 1. And for the largest group of potential PGD patients, those suffering from translocations, it may not be possible to establish a pregnancy at all without PGD, making such an alternative strategy irrelevant.

Hence, as Pickering, Polidoropoulis, et al. point out, "couples who seek prevention of transmission of a genetic disorder by PGD rather than prenatal diagnosis are highly motivated to avoid the emotional burden of further 'reproductive roulette'" (2003, 82). Similarly, Handyside and Delhanty describe PGD as most suitable for couples who would prefer to "know from the beginning that any pregnancy should be unaffected" by a specific type of disorder, in order to avoid "the possibility of having to decide whether or not to terminate an established pregnancy diagnosed as affected at a later stage of gestation" (1997, 270), and this is also similar to the claim by Braude et al. (2002) that "couples who choose PGD need to be highly motivated, as the process is complicated, expensive and, in some cases, associated with a lower chance of having a healthy baby than conceiving conventionally" (2002, 943).

It is for these reasons, ironically, that couples referred for PGD are, in an important sense, strongly cautioned *against its use* both by clinical staff and by genetic counselors from the outset. In contrast to IVF, which has been publicly criticized as "overprescribed" by the HFEA, the predominant rhetoric of PGD is surprisingly dissuasive. This "are you ABSOLUTELY sure" tactic is protective of both patients and clinicians. At both clinics participating in this study there was a strong sense of the importance of potential PGD couples having given *very* thorough consideration to their choice, its drawbacks, and alternative possibilities. On her first visit to Leeds, Sarah Franklin recorded her surprise at the explicit attitude of clinicians, nurses, and genetic counselors that their duty was to present "the really bleak picture" of PGD to prospective patients, thereby "putting the fear of God into them":

> To my great relief, Karen, the PGD coordinator, "adopted" me this afternoon and took me to several meetings between genetic counsellors and prospective PGD patients. I am amazed how much failure is emphasised, how much they go on about how many things can go wrong, how it is all very early days for the technique and for the clinic, and how complicated all of the procedures are. They always mention an earlier misdiagnosis, in one of the clinic's two attempts so far, producing what they refer to as a "50% possibility of misdiagnosis." They seem to want patients to be almost adamant they'll do nothing but PGD. Later a couple comes in for their third consultation. They are incredibly articulate and clear about their reasons for wanting PGD. The genetic counsellor says to them: "This technique is for people like you—people who know PGD is not the easy option, people who know this is a very difficult technique." (9 January 2001)

Once a couple has "really thought a lot about it" and has returned for a second or third consultation, a decision will be made if their case is warranted, and only then will an actual workup commence. Often a couple will already have had numerous genetic tests, but it can also be the case they have not had any. On the whole, the patients treated in London and referred to Leeds had lengthy medical and reproductive histories behind them, and came to the clinic at what the London PGD coordinator, Jenny Caller, describes as "toward the end of their reproductive lives."[18]

Through the Door

The question of how patients had "arrived" at PGD was the first thing couples were asked during interviews—and was always met with lengthy and upsetting replies. These "how we got to PGD" stories could begin far back in time, with an initial miscarriage, an affected birth, or a chance occurrence, such as reading a newspaper article about PGD. In discussing their "route" to PGD, couples often provided epic tales of hardship and struggle, in which a characteristic determination featured

---

[18] The most common sources of referral result either from the birth (and often premature death) of a child affected by a severe genetic disorder or from a sequence of miscarriages that are eventually linked either to a known genetic disease or to a chromosomal translocation. A third, and somewhat smaller, group of patients is comprised of couples who have never achieved a pregnancy and whose infertility is eventually determined to be genetic or chromosomal in origin.

prominently. "Getting to PGD" had often involved going through numerous painful experiences—not only of tragic events such as the death
of children or repeated miscarriages—but also of previous failed forms
of treatment, complicated family situations, and challenges to the couple's relationship. These interviews were consequently also often inspiring and very moving, as an individual or couple's ability to prevail over
their devastation to build a more positive relationship to the future
would be revealed alongside their vivid accounts of pain and loss.

These experiences are essential to understanding the process of "going
through PGD" and "moving on from PGD" (which are described in the
following chapters) because of the many aspects of "getting to PGD"
that are repeated, or revisited, during the process of going through it. In
addition, it is often the difficulty of "getting to PGD" that makes it appear an obvious, self-evident, or inevitable option once patients have arrived at its door. Similarly, having arrived at PGD as a way forward, it
may be difficult to "move on" if it is unsuccessful, for reasons that are
also made visible through what occurs "going through" treatment. A
prominent finding of this study, as in Franklin's earlier study of IVF
(1997), is the extent to which PGD patients described their decision to
undergo PGD in terms of "having no choice" or there being "no decision to be made" by the time they had reached their initial consultation
as prospective patients.[19] This sense of inevitability raises a number of issues about decision making and choice in the context of new genetic
medicine that are explored here by charting the personal histories and
experiences that led patients to PGD, and subsequently informed their
experiences of treatment.

At the same time, the sense of momentum or inevitability so strongly
evident in (and in some ways defining of) these narratives is also artifac-
tual: it is the predictable result of interviewing PGD patients rather than
people who decided, for one reason or another, not to pursue this form
of treatment. The reasons people decide to forgo PGD are likely to be as
varied as the reasons they opt for it, ranging from having become "spontaneously pregnant," to a decision to use an alternative method such as
gamete donation or prenatal genetic testing, to simply not having the
time or the resources to undergo such a demanding and expensive form
of treatment. However, while a complementary study of PGD "refusers"
would have added inestimable value to this study, it is both practically

---

[19] This language of "our only choice" is particularly striking in the context of PGD,
which has traditionally been described as an "alternative" to prenatal diagnosis. Thus, the
number of patients for whom PGD is, in fact, the "only choice" could be said to be small
and indeed is nonexistent if the alternatives of remaining childless, adopting, fostering, gamete donation, etc., are included. This theme is returned to at the end of this chapter, and
throughout this book.

and ethically complicated to contact people who have decided not to get back in touch with the clinic, and such a study thus remains one of the major areas awaiting further research.[20] Here, then, emphasis is placed on the factors that influence patients' decisions to pursue treatment, and on how these mesh with the desires of clinicians, geneticists, and other health professionals to offer improved treatment options. Together, these desires create a powerful momentum that plays a key role in the expansion of genetic medicine, and it is this trajectory of hope, belief in scientific progress, and a sense of obligation and responsibility that, when coupled with technological possibility, poses one of the most challenging questions raised about the social life of PGD.

Finding Out about PGD

Despite having been practiced in Britain for more than a decade, PGD is not a widely available treatment, and in 2001 it was still in many respects a very new technique. Consequently, the twenty-three patients who agreed to be interviewed between 2001 and 2002 comprise what could be described as its "pioneer" generation, in the sense that anthropologist Rayna Rapp uses the term "moral pioneering" to characterize the difficult decision-making processes, in uncharted territory, that beset the technique of amniocentesis. As Rapp writes of the women she interviewed in New York City during the 1990s:

> Women are both constrained and empowered through technologies like amniocentesis to serve as our contemporary pioneers. At once held accountable at the individual level for a cascade of broadly social factors which shape the health outcome of each pregnancy, and individually empowered to decide whether and when there are limits on voluntary parenthood, women offered amniocentesis are also philosophers and gatekeepers of the limits of who may join our current communities. (Rapp 1999, 318)

The couples undergoing PGD interviewed for this study are also pioneers in an historical sense, as many of them were among the first patient cohorts to undergo treatment as it became more widely available in Britain.[21] Finding out about PGD was, for them, an often lengthy and

---

[20] A grant proposal for a successor project looking at why some people choose the route of adoption rather than assisted conception was declined largely on the grounds that the two groups are not readily comparable (Franklin, McNeil, and Roberts 2002).

[21] It is important, too, to note the importance of the "pioneer" analogy for an American population, which limits, in some respects, its "translatability" to Britain. For example, the settler ethos of hardship and self-creation through being shorn of an identity described in

haphazard process, frequently beset by delays, detours, and "chance" events, such as hearing about PGD on a radio program or meeting a particularly knowledgeable clinician.

Increasingly, this is less often the case, as more and more potential PGD patients are identified and informed of PGD as an option earlier in their reproductive histories. As Jenny Caller, the PGD nurse coordinator at St. Thomas' clinic, describes this change:

> Initially we were seeing large numbers of couples who had been through quite a lot, and were nearing the end of their reproductive lives. These patients tended to be very well-informed about the procedure by the time they got to us, and they knew what was involved. Now that PGD is being recommended much more often, and has become better known, we are seeing a greater proportion of patients who know very little about it, and are embarking on treatment much younger, closer to the beginning of their reproductive lives.

Jenny's observation is reflected in several interviews for this study, which include some of the first generation of patients to undergo PGD with the Guy's and St. Thomas' team. This group could thus be described as transitional, in the sense that later generations of potential PGD patients will comprise a population more likely to have been referred directly to this technique.

Chance Encounters

Trish and her partner Ian were interviewed in London shortly following their decision to undergo a second course of PGD treatment at St. Thomas'. Now in her late thirties, Trish had undergone a lengthy process of finding out about PGD that is typical of its "pioneer" generation. Following two miscarriages, she underwent genetic testing and discovered that her mother was a carrier of a translocation. Trish describes herself as "very lucky" to have been referred by her gynecologist to a prominent clinical genetics unit for testing, where her translocation was discovered. As she recalls:

Frederick Jackson Turner's famous frontier thesis is explicitly about not being European (having one's "back" to the "mother country"). In this sense it is only a New World frontier that can create "character" in the Turnerian sense. The European frontier was more about living with the neighbors one "faces" or "fronts" (as in *frontiére*—to face or front). Another "genetic frontier thesis" might suggest that the problem for British PGD patients is not so much innovation as duty.

Right, well after two miscarriages, I was very lucky that my gynaecologist decided, er, because of my age, and he knew me very well, not to wait for three miscarriages, and he decided to run all the tests and things came through in dribs and drabs: I had a high prolactin level, um, I had various other things, but the main thing was this chromosome imbalance. He didn't know very much about it, so he sent me [to hospital X] for genetic counselling.

Although the discovery of a translocation led to a workup on Trish's family and thus to the discovery that her mother was a carrier, this process of discovery was, in Trish's terms, "very confusing":

And it was after the meeting at [hospital X] that it was explained to me, and I decided, ah, right, I need to find out more about this. And they didn't initially tell me about PGD, they said they wanted to do a work-up on my family, um, a sort of, um, er, genetic mapping. I think that's what they call it? And as it happened, my mother has exactly the same [translocation] as a carrier, um, but she has four children and, as far as she knows, no miscarriages. So that was sort of very confusing for me.

Since neither her gynecologist nor the genetic counselor had said anything about PGD, Trish began to undertake research herself on the Internet:

So basically, I got on the Internet and I was just feeding in "chromosomal inversions," *anything*, just trying to find out as much information as I could. And that's when a page came up and I think the organisation's called DEBRA.

Listed on the DEBRA (Dystrophic Epidermolysis Bullosa Research Association) web page were two St. Thomas' staff then in the process of developing the first PGD protocols. As Trish describes her reaction:

So all of a sudden *it just hit me*. [To Ian] Do you remember? (mm) I came down and I was [laugh]. . . . Ian was trying to read his paper, and I *threw* this page in front of him and, um, it just seemed to be the thing that I needed. It was, you know how you *grasp* at things and I, I just felt this is, this is the way forward. Um, so I tried to make contact with St Thomas' but I kept getting the answer machine.

The sequence Trish describes—of being worried about her age, feeling "lucky" to be referred for genetic testing, receiving a positive diagnosis as a carrier of a chromosomal translocation, which is confirmed within her family but leaves her feeling "very confused" so that, in the end, it is her own efforts using the Internet that lead her to the PGD clinic at St. Thomas'—typically combines several elements of patients' complicated

111

journeys to PGD. Moreover, Trish and Ian's journey was far from over. Unable to contact St. Thomas' immediately, she had "various conversations" with two of the other clinics licensed to provide PGD, was given an appointment at one of them that was subsequently canceled ("which obviously I was upset about"), and then shortly afterward became pregnant again, "so all ideas of PGD were shelved."

Late in 1999, following her third miscarriage, Trish decided to try to contact St. Thomas' again, in a state of "desperation":

> In the meantime I'd had an over-active thyroid gland, I'd had this pituitary tumour from the prolactin, and I was in a bit of a mess. I, I, I just felt quite *desperate*, really low about the whole thing, and it was almost as if I was just a punch-bag at this point for anything that could be thrown at me. And it was after Christmas, [to Ian] wasn't it, that we started to think let's, let's try St Thomas'. And I'd had a telephone conversation with them, um, but I got the impression that they were in a bit of a state of flux, so I'd sort of let St Thomas' go but it was, it was just, I don't know, luck, um, it was just synchronicity, I don't know, but I phoned them, and I had such a lovely response from St Thomas', such a positive response, and we went to see the doctor, and, um, well he was our first proper appointment.

In this passage, Trish's references to feeling in "a bit of a mess," "quite desperate," and "just a punch-bag" depict her frustration and yet also demonstrate her determination to persevere. In contrast, her repeated depiction of herself as "lucky" and her emphasis on the serendipity of events that led her back to St. Thomas' convey a sense of random, or chance, occurrences. Later in the interview, Ian describes Trish as having intense determination: "she's like a terrier, really, she goes for something and sort of pursues it, so I mean its very admirable," and yet he too adds that "obviously it was a bit of luck to a certain extent." This view of "making your own luck" belongs very much to the world of cutting-edge reproductive medicine such as PGD and remains a central feature of the experience of undergoing IVF. The theme of "creating chances" is repeatedly encountered in PGD and is explored in greater depth in chapter 4. Getting to PGD, in Trish and Ian's case, involved numerous "chance encounters" with clinicians, online patient information groups such as DEBRA, and the St. Thomas' clinic itself, where Trish was initially thwarted by a busy phone line, and a second time by the impression the clinic was "in flux," before being "third time lucky" attending her "first proper appointment."

Trish and Ian's eventual arrival at the St. Thomas' clinic, then, was the result of "lucky," "serendipitous," and "chance" encounters while at the

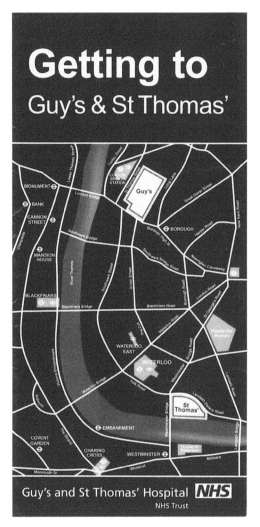

FIGURE 3.2. Finding their way to PGD often brought patients along lengthy, un-expected, and surprising routes.

same time it was also very overdetermined, in the sense that several different avenues pointed toward PGD from the outset. From the moment of discovery of a genetic cause for her serial miscarriages, Trish was on a trajectory likely to point toward PGD—if not immediately then soon, and in the future with increasing inevitability for couples in her and Ian's situation. For Trish, there was a significant gap between her own determination to succeed and the lack of awareness about PGD among the

113

numerous medical staff she encountered over several years of repeated miscarriages. While praising the gynecologist who initially referred her for a genetic workup at hospital X ("who's actually quite eminent"), Trish expresses regret that although "he seems to have his eye on the ball" he nonetheless "had no idea" about PGD and, when informed of her positive test for a translocation, "didn't seem to know much about that either." Indeed, Trish adds, "when I wrote back to him and asked about PGD, what do you know about that? he really, he'd sort of heard it somewhere in the fog."

Trish describes this contrast by comparing herself to another, hypothetical and less "obsessive," patient in her position:

> And the point I'm trying to get at, if, if, if a woman who (a) didn't have access to the Internet, and perhaps wasn't as obsessive as I am, about finding out the information, say someone who perhaps wasn't particularly well-educated, had just had this result, um, she may have just sat back and said, well there's nothing that can be done for me. I'll just keep trying, myself.

Significantly, Trish and Ian's "first proper appointment" was tremendously positive. "They did a work-up on our bloods, and they were so positive! And then it just proceeded from there." Despite failing their first cycle of PGD, Trish describes a continuing feeling of appreciation of the clinic, and being buoyed along by the momentum of treatment, and her hope for eventual success:

> What I really, *really* appreciated about it, as opposed to becoming pregnant naturally, and, um, then losing and feeling in a vacuum and feeling lost and lonely, and all those other feelings, because you've got this, almost day-to-day interaction with—which is what people must experience as well with IVF—the hospital visits, the scans, the sniffing, and the injections, um, the contact with the hospital keeps you going, it makes you feel more *alive*.

Like earlier generations of IVF patients, who were more likely to arrive at treatment later in their reproductive lives, having often experienced frustration with what they later perceived to have been inadequate or even incompetent primary care, Trish expresses relief at being cared for by a positive and supportive team. She adds that it is "a relief to have someone sort of directing you, you're going in a particular direction." And she notes that "once you start being picked up by the system, it develops a momentum of its own. . . . I feel that once you seemed to come on board, you were carried by a lot of it."

Trish's feelings of being "picked up by the system," taken "on board," supported on a day-to-day basis by her interactions with the clinic, and given a direction are summarized in her account of PGD treatment having "a momentum of its own." This sense of momentum is all the more understandable given what she and her partner have been through in the past, the lengths to which they have gone to find treatment, and the desperation she felt at her lowest point, just prior to her first "proper appointment." "Getting to PGD" is thus a two-sided process for many couples. It is precisely the difficulty of "getting somewhere" that can make arrival seem both "lucky" and the result of intense determination.

This determination is an important characteristic of PGD users in several senses, for, by definition, such people are exceptional. Combined with the fact that the clinicians who offer PGD are also exceptional, and often exceptionally driven, a potent combination of will and possibility emerges, which could be described as one of the major points of conjunction that creates a sense of momentum not only for patients, or the clinicians who treat them, but for the wider public, who may view techniques such as PGD with concern. As another interviewee, Steven, describes his first impression of PGD: "it was there, it was possible, and so we decided to go for it." Importantly, this powerful meeting—between the desires of patients eager and determined to find treatment and the intense dedication of clinicians who are making every effort to provide it—creates a highly emotional encounter, which is discussed further in chapter 4.

While Trish and Ian's case reveals many aspects of the road traveled by many patients to PGD through numerous ups and downs, "lucky" and disappointing encounters, and the sense of a long path through confusion leading to relief when finally they are "taken on board," their experience of PGD was very different from that described by interviewees who had witnessed the death of a child from a severe, terminal genetic disease. In such a situation, issues of loss and responsibility both simplify and complicate reproductive decision making, and comprise another important component of "getting to PGD."

Losing a Child

Anne and Daniel were interviewed in their pretty bungalow in the south of England while having a break from PGD after their third failed attempt. The mantelpiece of their tidy, carefully decorated sitting room was crowded with pictures of family and friends, and on top of their TV stood a large studio portrait of a very pretty baby, at about six months. This was of Chloe, their daughter who had died at eleven months of type 1 spinal muscular atrophy (SMA). As Anne and Daniel described Chloe's initial diagnosis:

115

DANIEL: When Chloe was diagnosed, *well we didn't know*, we had a hunch that we were going to this meeting, and we were going to be told . . .

ANNE: We thought the *worst case* scenario . . . [*speaking together*]

DANIEL: We didn't know that there was something wrong with her. We had a, you know, a fair idea that there was, something was coming, um, and we sort of said, right, all right, what's the *worst thing* that can happen—she might be in a *wheelchair*. *Great!* We can live with *that*!

ANNE: We live in a *bungalow*!

DANIEL: We live in a bungalow with no stairs! We'll put in a *ramp*, you know, we'll get her a little bloody powerchair, and you know . . .

ANNE: We'll be *fine*!

DANIEL: The world's *rosy* again, kind of thing! You know, we can *live* with that! Um, and that's not like—it wouldn't have been as much, it wouldn't have been, we would obviously have been, sort of heart-broken and shocked, and everything, but we would have just gotten on with our lives after finding out!

ANNE: I think we would have *eventually* readjusted, we would have just, um, but it was different to that. Because we walked into the room, and, er somebody said I've got some very difficult news for you—your daughter is not going to live past her first birthday. And I mean that's a bit *different* to somebody telling you that, you know, she's got a disability and she's going to be in a *wheelchair*. [*pause*]

SF: And you can't take in a statement like that anyway, can you?

ANNE: No. No. I often say to people now, um, I can't remember anything about the journey. Or how we got home at all that day.

SF: Was Chloe with you?

ANNE: Yeah, yeah, she was. That's what—I can remember being so hysterical that I had to hand her to somebody. A complete stranger! Cause I was, I just couldn't hold her. I was hysterical.

Chloe's death completely changed Anne and Daniel's understanding of their reproductive futures, by introducing a powerful new sense of obligation toward their potential offspring—to prevent their ever suffering as Chloe had. Speaking of how Chloe's life and death changed their views about having children in the future, Anne is emphatic:

Well yeah, I mean she's influenced us completely. Because, you know, we just couldn't go through with another baby with SMA type 1, we just couldn't do it. Um, we couldn't do it ourselves, but we couldn't do it to one of our children, or our family. It would just be a definite no.

Daniel explains further:

> I think its quite hard for people to understand, cos people think like that we regret having Chloe. But it's a totally different thing. We didn't know that Chloe was affected with SMA, but we would know that, you know, if we got pregnant naturally and, you know, we had another child, we would know that we'd, we'd had the choice, and we've produced another child with SMA!

To look at such a child, Anne explains, would inevitably involve a deep sense of regret, of irresponsibility, and of culpability. The sense of knowing there was something that could have been done to prevent such suffering, and the prospect of knowing that choice had been forgone, created the "definite no" to having children in the future without going through PGD.

Remembering Chloe's final week, her quality of life, and how much they loved her, Anne and Daniel describe the burden of responsibility that knowledge of a potentially lethal inheritance to their children imposed on them:

> ANNE: I think for me, it, it's, you know, just thinking of sort of Chloe in her final week, um, the final days of her life and it's thinking of doing that again, looking at that baby, and thinking I could have done something about this, I knew this was going—, could have happened and I could have changed it. Whereas with Chloe, we didn't know anything. I mean, we always said that the eleven months we had with Chloe were—I wouldn't change them for anything! You know, we're so grateful that we had her and, um, that the quality of life that she had was so good, and um, I mean, we loved it so much that we would want to do it all over again! But I wouldn't choose to relive those eleven months, I wouldn't choose to do that to another baby. No. That's the difference, isn't it?
>
> DANIEL: That's the difference for us.

An option for couples such as Anne and Daniel would be to become pregnant "naturally" and undergo chorionic villus sampling (CVS) or amniocentesis in order to diagnose SMA prenatally. In both the clinics where couples were observed being counseled about PGD, it was repeatedly emphasized that such a method would be likely to produce a healthy child more quickly. Since PGD is time-consuming, expensive, and relies on IVF, it is not the most efficient route to pregnancy, or to genetic diagnosis. However, the prospect of either CVS or amniocentesis raises difficult issues for couples such as Anne and Daniel, who have experienced the death of a very young child. As Anne explains:

I mean the one thing we realised was that any pregnancy was going to be difficult, and even getting pregnant through the normal routes, we were going to be faced with a CVS test at 10 weeks, and some difficult choices if the baby was found to be affected again. I mean the one thing we said sort of straightaway was that, you know, there's no way we could put another of our children through Type 1 SMA. It was different with Chloe, because we didn't know we were carriers until we had her, but now we do know, and there's no way we can have two of our babies dying in hospital, we just can't do that.

Like many other PGD patients, Anne discovered PGD "by chance," in her case while reading a book by Robert Winston she had consulted in order to find out more about CVS. "I was just sort of reading and all of a sudden I came across it and started to read a bit faster!" She and Daniel subsequently researched PGD on the Internet and found out about the St. Thomas' clinic, to which they wrote a letter explaining their situation. Describing "what initially tempted us to PGD," Anne says it was "the fact that we, you know, we're not talking about a nearly fully developed fetus or whatever, you know, we're just talking about a few cells." Whereas a PGD pregnancy would be treated as "just a normal pregnancy," the prospect of not knowing for several weeks if the pregnancy would continue, and facing high (one in four) odds of a positive diagnosis for SMA seemed "quite daunting" for Anne and Daniel. As Anne explains:

I think being pregnant, sort of naturally, as well and the sort of trauma from finding out you're pregnant, till sort of getting to 10 weeks and getting to the test, not knowing whether you're going to go ahead with the pregnancy or not, that seems quite daunting to us really, and although PGD is in our experience, has been very emotionally demanding, I mean it's not easy, but, um, I mean certainly it seemed like I suppose if you got pregnant through PGD, you'd be pregnant, and then it would be treated as just a normal pregnancy. Whereas having those sort of five or six weeks before you even go for a test, and then having all those difficult choices, seems, it [PGD] seemed more appealing at the time.

Yet, while Chloe's death made Anne and Daniel feel more daunted by an uncertain pregnancy and the possibility of termination, they also felt their views toward termination had been changed by losing her. In the past, before they knew they were carriers of SMA, they "would not even have considered the possibility of termination." Now, even if they knew they were having a child who would be disabled, they would not terminate. But Chloe had been more than just disabled. The difference," as

Anne explains, "is that Type 1 babies don't live to see their first birthday. I mean there is just no sort of future at all in it, really, is there?"

With funding from their local health authority, Anne and Daniel decided to go ahead with their first cycle of PGD, despite having had "the blackest picture" of it painted for them at their first consultation in London. As Daniel explains:

> We came away thinking "Crikey, that doesn't sound very good," but we were still determined to do it, weren't we? And give it a go, and I think throughout it all, really, I think we always thought, well we have nothing to lose.

Infertility, Translocations, and Misdiagnoses

As noted earlier, there are a number of reasons why, possibly in Britain in particular, PGD is clearly distinguished from infertility treatments. It was often explained to visiting television crews, to patients, and to scientific and policy-related meetings, that PGD patients are usually not infertile. "We firmly believe," Peter Braude stated in a PGD study day held at Guy's Hospital, "that PGD is not an arm of assisted conception, it is an arm of clinical genetics. Assisted conception is simply the technology by which you are able to do it" (May 2001). The reasons for this distinction are partly scientific and partly related to resources.[22] As the head of Britain's leading PGD clinic, Peter Braude wants to insist on the specificity of PGD in order that as many patients as possible receive public funding for their treatment.

The issue of PGD and infertility is somewhat more complicated by factors such as age, and by the effects of both chromosomal and genetic conditions (such as translocations and cystic fibrosis) on fertility. Among the couples interviewed for this study were a significant number who had experienced infertility. All these couples had sought, and some had actually undergone, conventional infertility treatments such as ovarian stimulation and IVF. In some cases, an underlying genetic or chromosomal component of their infertility had been picked up in standard testing before an IVF cycle, and the couple had been referred to a clinical genetics unit. In other cases, the couples had actually undergone IVF, with the genetic condition remaining undiscovered over the course of several failed treatments. For those couples whose heritable conditions were

[22] The reasons for associating PGD with genetic disease can also be interpreted in the context of Britain's historical debate about the moral importance of scientific progress to alleviate the suffering caused by serious genetic disease, as noted earlier, in the first section of this chapter.

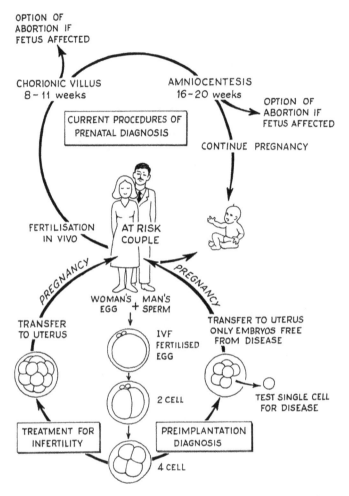

OPTION OF
ABORTION IF
FETUS AFFECTED

CHORIONIC VILLUS
8 – 11 weeks

AMNIOCENTESIS
16 – 20 weeks

OPTION OF
ABORTION IF
FETUS AFFECTED

CURRENT PROCEDURES OF
PRENATAL DIAGNOSIS

CONTINUE PREGNANCY

FERTILISATION
IN VIVO

AT RISK
COUPLE

PREGNANCY

PREGNANCY

WOMAN'S
EGG
+
MAN'S
SPERM

TRANSFER
TO UTERUS

IVF
FERTILISED
EGG

TRANSFER TO UTERUS
ONLY EMBRYOS FREE
FROM DISEASE

2 CELL

TEST SINGLE CELL
FOR DISEASE

TREATMENT FOR
INFERTILITY

PREIMPLANTATION
DIAGNOSIS

4 CELL

FIGURE 3.3. Options for couples at risk as depicted in one of Marilyn Monk's visual diagrams of PGD.

diagnosed (and one must assume that many more were/are not), some had undertaken additional assisted conception procedures, such as sperm or egg donation. In all these cases, attempts to use non-PGD-related assisted conception had failed, and were described as frustrating, depleting, and emotionally difficult.

For a number of interviewees, routine testing in preparation for an IVF cycle had led to the discovery of a genetic translocation. Such a discovery indicates that standard IVF will be inadequate, as there will be no way to determine which embryos are affected without biopsying and assaying them (the normal grading of IVF embryos is visual). In the case of another couple interviewed, routine sperm testing led to the discovery of aspermia (no viable sperm in the ejaculate). Genetic testing found that the male partner had two cystic fibrosis (CF) genes and that the female partner had one CF gene. Again, this means that standard IVF is unsuitable, as embryos carrying CF genes may be transferred, and the disease itself impairs fertility.

Although early detection of the genetic problem was beneficial to some couples in preventing unnecessary IVF cycles, the experience of obtaining and dealing with a genetic diagnosis was also disturbing and difficult for others, as Rebecca and Steven's experience demonstrates.

A professional couple in their midthirties, Rebecca and Steven sought medical advice after failing to get pregnant after eleven months of trying. Concerned to get good-quality treatment as quickly as possible, they arranged an appointment at a local private clinic and were seen by a leading clinician. They were about to start their first cycle of standard IVF when results of routine tests came back with bad news.

> SF: How did it first come up that there was a translocation?
> REBECCA: I tell you that was awful! I remember that episode, because what they did was they said um right we can go ahead with taking all the samples they need, and we could go ahead with IVF. And you've got to phone up when it's the first day of your period and by that time she said we'll have hopefully all the tests back, and we can, you know, you'll be on the cycle. And I phoned up and the nurse said "oh um, right, um no, there's a problem. You've got to come in and see Dr L." And I thought—"cause you've got yourself *so* built up to this you know—"this is it, this is our chance." And then they wouldn't tell us anything over the phone! . . . The next time we could see Dr L. was like a month later! And suddenly from being ready to start, nobody would tell us *anything*! I can't tell you! You think, "oh my god, *what* is going on?"

Following the diagnosis of a genetic disorder, Rebecca and Steven spent a year having widely spaced discussions with IVF clinicians and counselors, in which they were offered (in ways they experienced as quite directive) alternative solutions of adoption and sperm donation. Despite a long-term interest in adoption, and concerns about the emotional implications of donation for themselves as a couple, Rebecca and Steven undertook an IVF cycle using donor sperm. Rebecca describes this experience as "horrible" and poorly managed.

Both Rebecca and Steven describe a sense of time slipping away, and an increasing sense of frustration. Indeed, Rebecca comments that this period of waiting was for her more stressful than the eventual PGD procedure itself.

121

Oh god! The *stress* of—of going through this whole thing! When you're *actually* doing it it's fine, it's never bothered me. No problem. It's the *bloody* waiting! . . . We had this year in which *every* time we went in to see him he told us *another* bit of bad news! Which meant, you know, not even—you sort of condition yourself to thinking "right I'm going have to go through IVF, right OK," condition yourself to that, and then some other problem—it's like, oh! You know, there were sort of a few months between, and it's *layer* and *layer* of bad news. Kept building up, didn't it? (Steven: Yeah). I hated going into his office actually!

Finally this period ended when the IVF clinician wrote on Rebecca and Steven's behalf to all the existing PGD clinics, and St. Thomas' agreed to offer them treatment. Nevertheless, this period of waiting remains an unpleasant memory, of a lengthy period of "wasted time" during which the seriousness of their situation and the difficulty of obtaining suitable treatment only gradually became evident. As Steven recalls,

That was the *hardest* bit, I think, because we spent probably a year, just, just with him. Just literally talking about various bits and bobs and *nothing* seemed to be moving on. And he'd call us in and we'd go in and we'd spend an hour and pay our fifty pounds and we'd go away and we're no further on really! And there were a lot of tests going on then, and I think it became evident at the end of that year that ah, we had something that was more serious than just straight IVF or ICSI [intracytoplasmic sperm injection]. . . . What he then did is wrote to, I think, five different hospitals to say look this is the problem, this translocation that I've got, and [asking] whether any of them would take me on board, so to speak. And um, I think the most shocking thing was that quite a few wouldn't.

Although this period of waiting was difficult, other couples reported even greater retrospective frustration and disappointment, because they had undergone standard assisted conception treatment despite having a translocation (whether the translocation had been identified or not).[23]

In Belinda and Nicholas's case, two ectopic pregnancies and a miscarriage led to multiple attempts at increasingly invasive forms of assisted conception including ovarian stimulation, intrauterine insemination (IUI), and standard IVF. Belinda and Nicholas disagreed during their interview as to exactly when the genetic translocation had been diagnosed, but it was clear to Belinda at least that these IVF cycles had taken place

[23] For this study, three different couples were interviewed in which the female partner carried a genetic translocation but had engaged in multiple assisted conception and IVF cycles that seem, in retrospect, to have been a "waste of time."

in spite of the IVF team's knowledge of her genetic condition and the consequent unlikelihood of success. The clinic did not offer a PGD program and did not make a referral to a specialist unit. It was not until Belinda became pregnant with their fourth IVF attempt and then lost this pregnancy in an early miscarriage that this referral occurred.

> BELINDA: We'd done a number of IVF treatments with other hospitals, and we asked if there was any research into, you know, isolating the bad eggs, 'cause I had a chromosomal translocation.
> NICHOLAS: Although you didn't know that then.
> BELINDA: And she said that there wasn't much research into it, but that was something that was being worked on, and would be something that would be available at some point, but it's not something they did. And after three or four failures, they said that they couldn't treat me anymore because it wouldn't be fair. And they referred us to Guy's and St Thomas' because they had the research facilities and they were doing PGD research. And, and it was through, Dr W.—she wrote to St Thomas' and they took us on board to the programme, so that's in a nutshell how we came across it. We have differing opinions of when and how we were told, but that is exactly how it was.
> SF: So those cycles of IVF, how far did they get? You said you did four?
> BELINDA: The last one I did there, it took, and then I lost it, after a couple of weeks and that, they said that was down to the fact that it was probably a bad embryo, um, which had a problem. And it was at that point they said it wasn't fair to keep *trying*, pot luck, to see if I'd get pregnant with a good one.

Both Belinda and Nicholas felt it was problematic that the clinic continued to offer them treatment when it was so unlikely to succeed, and despite their having explicitly requested genetic tests. Here again, as for so many PGD patients, the already arduous path through IVF was made more difficult by the intersection of reproductive and genetic factors affecting fertility. As Nicholas points out, the odds were doubly stacked against them.

> Well I didn't think that [genetic testing] . . . happened until after a couple of [failed] attempts. . . . And I also think that it was at our suggestion that the tests were done and that even *after* that, they went ahead with the last treatment and then they pulled away. So they *were* cognisant for either *three* or one IVF treatments, as to Belinda's chromosomal, er, abnormality, and still went ahead with it. . . . I think if they'd been realistic about, you know, statistically, one in

123

four of Belinda's eggs is perfect. Divide that by the chance of actually getting pregnant through IVF, you've got about a 5% hit rate, if you're *lucky*! And we now know through the treatment at St Thomas' that, um, *often* the poor quality cells manifest themselves at a greater percentage than 25.

Given this difficult history of unsuccessful infertility treatment, it is not surprising that Belinda reports a strong sense of relief at being referred to the St. Thomas' PGD clinic, which she describes as being "passed into the hands" of a specialist team. Having reluctantly come to believe that "there was nothing that could be done" about the translocation, "the relief," Belinda says, "was that we were being passed into the hands where they had research that they could actually isolate it!" For Belinda, this was "what we'd been hoping for. Our only chance, really." Nicholas shared this sentiment, but adds an important proviso, with which Belinda concurs: "But at the same time, obviously, even though we'd identified the problem, it was quite a big problem." Arrival at the PGD threshold for Belinda and Nicholas, then, had quite a different tone than for couples such as Ann and Daniel, or Trish and Ian. While the PGD clinic seemed to offer hope, their long experience of reproductive failure and "wasted" treatment meant that they did not have overly high expectations of what could be achieved. PGD offered a welcome and increased level of control, but no certainty. All the same, Belinda describes these new prospects as exciting and reassuring:

> When they explained it to us, I have to say I felt *excited*, because I really thought that this was it, you know, this was the chance. Whereas before it was just a stab in the dark, this time it just felt like it was being controlled properly. You just felt that you were in proper hands instead of leaving it all to, to, to chance. And of course there's obviously a *large* element of chance there, but it just *felt* that it was being more controlled.

The language Belinda uses, of being excited and of finally being "in proper hands," invokes two essential contrasts that are repeated in many IVF narratives, as well as PGD stories. The first important contrast is temporal: *before*, it was "just a stab in the dark," but *now*, "it was being controlled properly." This contrast between the "dark" of "before" and the relief of feeling more in control invokes the traditional view of modern science as cumulative and progressive, while also conveying Belinda and Nicholas' sense of arrival at the clinic: finally they have arrived somewhere *where something can be done*. The second contrast, between chance and control, adds what is often referred to as a paradoxical dimension of PGD in that intensification of the power to diagnose is also, at another level, amplification of pathology.[24] At one level, chance and control are opposed: instead of "leaving it all to chance," there is the option of "being controlled properly." However, since arrival at PGD confirms the presence of more serious disease, the prospect of "greater control" is double-edged (possibly more control but only in relation to a worse set of prospects). What is being described here in the language of hope is a *feeling*. Instead of the situation being out of control, or out of hand, Belinda *felt* there was more control "in proper hands." Indeed, this was what made "the chance" of being able to undergo PGD so exciting, that "it just *felt* that it was being more controlled."

As Belinda's comments reveal, the language of hope, chance, and choice is both inherently paradoxical and highly emotional. For example, the expression "chance" conveys both a sense of opportunity ("the chance") and sense of an unknown outcome, as Belinda confirms when she acknowledges that, while she may not be "leaving it *all* to chance," "there's obviously [still] a *large* element of chance there." In this sense, chance and control are not so much opposed as relativized—there is a better chance (opportunity) because there are fewer chance factors (less randomness). Significantly, this *emotional feeling*—of excitement, and relief, about the prospect of more control—frequently characterizes the use of language to describe a path ahead in which neither certainty nor control feature prominently, and hope is the only semisolid ground on offer.

A Perfectly Valid Clinical Option

As these cases demonstrate, couples commonly arrive at the PGD clinic with complex reproductive histories, often involving significant losses, through which they have acquired greater understanding of their genetics and fertility. Their relationships to such intimate and profound issues are inevitably highly emotionally charged, and a significant amount of self-education about various alternatives has often increased their sense of determination to seek out the newest treatments. This determination is often strengthened by a deep sense of conviction or obligation to prevent further harm—to potential or existing offspring, to themselves, or to their families and communities—from the consequences of serious genetic disease. They have often been struggling "against the odds" for some time—more than a decade in a number of cases encountered during this study. For couples such as these *who decide to engage*

---

[24] See Armstrong (1994, 2001) for an early discussion of "surveillance medicine," and see Nelkin and Tancredi 1989.

125

with PGD, past experiences act as a powerful motivating force—leading to PGD as *one of the only clear and obvious choices* in the midst of the trauma of failure and misdirection.

Clinicians, nurses, and genetic counselors involved in PGD, then, often encounter determined, well-informed, and articulate professional couples.[25] While the staff of the PGD team may be keen to offer PGD to such couples, there is also a strong sense of shared obligation to ensure that couples are not holding overly optimistic views of what PGD will do for them. This produces a complicated process whereby the desire by potential PGD patients for the choice PGD can offer must be judged as to its suitability, in order to fulfill a duty of care neither to compound previous disappointments nor to raise unrealistic hopes of what PGD can offer. This process, in which the health professional becomes both an interlocutor and a potential barrier, "in between" what PGD can offer and the situation it is sought after to relieve, is the "bottom line" of the PGD consultation process.

The challenge this situation presents symbolizes the whole "question of PGD," as Peter Braude demonstrated at one of the clinic's PGD study days in April 2002. The audience of approximately sixty participants was largely composed of genetic counselors, geneticists, nurses, embryologists, and other specialists with an interest in PGD, or the possibility of setting up clinics in their own centers.[26] PGD study days provide the Guy's and St. Thomas' team with an opportunity to educate other health professionals about PGD, so that suitable couples are referred for PGD and the benefits of the technique can be better understood. Among other purposes, study days are a step toward a future in which more potential PGD patients are properly identified, referred, and counseled about PGD, and in which fewer patients who are unsuitable for PGD, or inadequately prepared for its demands, will pointlessly consume scarce clinical resources, such as consultation time.

Peter Braude's introductory PowerPoint begins with a series of questions: "So who might need PGD? Why bother about this? Surely prenatal diagnosis is adequate?" To illustrate why in some cases it is not, Peter uses the example of SMA—much in the same way Bernadette Modell's cases were presented by Anne McLaren to draw attention to the particular difficulties faced by parents who desire to "know from the outset that their pregnancy is unaffected by a specific disorder."

---

[25] Of course, and as mentioned elsewhere, these are not the only kinds of couples who consider PGD: other couples at both London and Leeds were neither professional nor middle-class, and couples were interviewed who do not fit into the determined, well-informed, or articulate category of PGD patients.

[26] Some health professionals attend such meetings as a means of maintaining a required level of continuing professional education, for which they receive formal accreditation.

Using a slide of a baby with SMA, whose limp body clearly demonstrates severe muscle wastage, Peter provides a brief but sobering clinical description:

> Here is a child who is under the age of one. You can see it's very floppy, it can't hold its head up. You can see the ways its legs are here, and this child will almost invariably die before it reaches eighteen months, usually within a year. And in all the cases of couples we have seen with children with this disease [the children] have died within a year.

He explains that one of his patients who had a child who died of SMA then went on to have terminations of two further pregnancies. After these two terminations, and the loss of her first child, "she turned around to us and said no more, no more. I don't want any more terminations. *Haven't you got something else?*"

This position "in between" the patient's question, "Haven't you got something else?" and PGD is at once "obvious" and difficult to explain—once again reproducing the defining dilemma of PGD, that it is a perfectly valid clinical option but not an easy choice, and not for everyone. As Peter knows from considerable experience, it is also not necessarily an easy topic to discuss with patients, because their desire to find the "right" treatment may lead them to overestimate the chances of success with PGD. As he continues, "And to actually talk to a couple like this and say well you could go ahead and try again because realistically your dice have been set to zero again and you only have a one in four chance, she doesn't believe you." Speaking on behalf of the woman patient, he repeats, "she does *not* believe you and she wants an alternative."

In other words, many patients are highly motivated to find an alternative to repeated terminations—*even if becoming "spontaneously pregnant" followed by prenatal screening is the advice they receive from specialists.*[27] Indeed, as Richard Penketh and Anne McLaren pointed out in their 1987 review of the prospects for PGD, patient groups for diseases such as thalassemia formally petitioned the government for research on PGD during public debate of the Human Fertilisation and Embryology Bill. The position Peter Braude is describing is historically established as the defining logic of PGD provision in the United Kingdom:

---

[27] "Spontaneously pregnant," like "standard IVF," is a vintage term of the era of reproductive assistance, in which, by definition, what came before is also marked by this assistance, becoming unassisted. "Spontaneously pregnant" denotes unassisted conception in a manner that makes invisible everything that is not spontaneous about becoming pregnant, because the referent is technological.

These are the couples who are coming along to us with all these kinds of stories and saying please can you give us an alternative. And I think these are perfectly valid reasons for wanting to do PGD.

Peter is on equally well-trodden territory summarizing the options for the couples who come to him wanting PGD, such as the parents of the SMA baby he showed earlier:

So when you look at what couples have on option if they've got genetic disorders, they can continue to roll the dice and see what happens, and some couples decide to do that if they are young enough. Some decide they can cope with prenatal diagnosis and termination of pregnancy. Others may decide that they will want to swap one of the partners who's a carrier, so . . . they could decide to use sperm donation . . . or more difficultly . . . ovum donation. . . . A number of couples say that they'll try and adopt because they will start afresh. And unfortunately many couples decide to remain childless. They don't want to take the risks and they don't want to pass this on down generations. *What we now can provide them with is the option of PGD.*

Significantly, PGD is presented here not as a solution *but as an option*, and indeed as but one addition to a long list of options for people with genetic problems. The case for PGD is not being made, as it once was for IVF, in terms of providing "miracle babies" for desperate childless couples who have nowhere else to go. In place of the confident superlatives of IVF are the extreme caution and reservation of the PGD pitch. The case for PGD is not even entirely medical, as Peter emphasizes in his account of the superior statistical odds of prenatal, rather than preimplantation, genetic testing. Indeed the case for PGD is almost being depicted as an option that is *primarily undertaken for emotional reasons*: because a couple cannot tolerate the emotional distress of repeated terminations, or because they need to feel they are doing something "more" than allowing the genetic dice to roll on *unaided by any helping hands.*

The picture, then, following the "cautious optimism" toward PGD in the 1990s, might be called one of *guarded expansion* at the dawn of its second decade. Not all genetic diseases are suitable for PGD, and, even when they are, the technique is offering, at best only the same "odds," one in four, of having a child that is neither a carrier nor homozygous for a single gene disorder as couples could achieve "spontaneously." As in Belinda's account earlier, this "dicey" aspect of PGD is also depicted by Peter Braude in terms of "chance" and "chances," or what is elsewhere referred to as "reproductive roulette" (Pickering, Polidoropoulis, et al. 2003, 82). But since the actual odds of the gamble are not what is being changed by PGD, the treatment being offered is for something else. It is in this way that PGD remains, and is acknowledged to be, a treatment for grief, by offering the medicine of hope—in the form of at least an element of added control but, perhaps more, the comfort of being, at last, in the hands of someone who can *do something*.

Not an Easy Choice

The position of being "in between" the chance PGD offers and the chanciness of its success is one of the most demanding aspects of PGD for professionals in this field. As Jenny Caller, the St. Thomas' PGD coordinator, patiently explained in a lengthy interview, no one wants PGD to be seen as a "magic bullet" for families with a genetic condition. On the contrary, what couples need to understand is that "it's not a simple treatment—it's very complicated, it's hard, and it's not always successful."

Fiona Robson, a genetic counselor at the LGI, similarly defines the most important part of her job as communicating the "reality" of PGD to hopeful couples who are potential patients. As she makes clear in her reference to the "*much* better chance of having an unaffected child by conventional methods," her job is to make sure couples have very thoroughly considered the alternatives to PGD. Like Jenny, she also considers this one of the most difficult parts of her job:

> To my mind, that is the difficulty in the whole technique of, um, not being downhearted about it or negative about it, *but to try to place the realism*. In fact, actually, they've got a *much* better chance of having an *unaffected* child by conventional methods, but we'll have to confront termination with that. It may be that it'll work first time *fine*, but maybe it won't. But, um, if you tease out the fact that *the most important thing of all is that they want to have a baby*, well . . . then it's going to be much more likely that they get it through the natural forms.

Like Jenny, Fiona is interested in supporting individuals to make the best decisions about their reproductive lives, and is at pains to explain the complexity of this option:

> This is *really* hard work. You don't want to dampen somebody's hopes. Some children are born now [through PGD], but you know it's—. For some families it's absolutely right. Also, even, some people just need to pursue it and then fail and, and then they can start again with something else. They just have to get that part out of their systems, as it were, and try. And we really *really* appreciate that but,

[sigh] I think also, um, you, *you have to somehow place the reality in front of them*. And try and get them to *examine* maybe some of their beliefs that they have, of why they could never think about having a termination of pregnancy.

What the core elements of Fiona's and Jenny's messages repeatedly stress is the difficult work of assessing what Jenny refers to as potential patients' abilities to get to grips with "the reality" or "the realism" of their chances—which is crucial to their decision as to whether PGD is an appropriate path. Similarly, the view Fiona expresses that this reality testing is "really hard work" indexes again the exhausting "emotion work" of PGD.

Realistic Chances

In the first section of this chapter, some of the divergent patterns of PGD usage internationally suggest that PGD is always to some extent a nationally specific technology. Thus, for example, PGD in Britain is deliberately less associated with fertility enhancement technologies than elsewhere, and its association with clinical genetics is consistent with the historical role it has played in establishing greater public support for innovation in the field of reproductive biomedicine. This chapter, as elsewhere in this book, also suggests that the view of PGD as a technology that has made "slow progress" because of both its technical complexity and the unpredictability of embryonic development (as well as other factors, including the range of specialisms it brings together) is possibly more characteristic of the way it is seen in Britain than elsewhere.[28]

Nonetheless, while it is inevitably difficult and hazardous to attempt to generalize about PGD in what is not only a complicated international context but in many ways an equally diverse national one, the data from this chapter add support to the observation made elsewhere that, from the beginning, *part of the effort to promote PGD has been to restrict its use*. Moreover, while it is equally difficult to say in any simple sense what PGD "is," since there are many "PGDs," it can be noted that for every version of "successful" PGD there are many more precautionary "failure" PGDs. In casual conversations, and also formal interviews with clinicians, nurses, PGD coordinators, and genetic counselors, a constant and consistent theme is that although PGD is a valid and necessary choice, it is not for everyone, is very difficult, and often fails. This find-

---

[28] Michael Mulkay, in his account of the expansion of assisted reproductive techniques and possibilities in Britain, describes a "cautious, gradual, almost imperceptible movement" into the future (1997, 154).

ing, that being pro-PGD seems almost to require the performance of be-ing "anti" it, points to an important paradox at the root of the PGD op-tion: while much of its progress has been driven by the intense emotional excitement about the prospects, or "chances," it offers, it is this same emotional intensity that is seen to be a potential source of harm.

These observations lead to others, especially regarding the intense emotionality of PGD. While potential PGD patients may be viewed somewhat warily by the clinical team and genetic counselors who fear that such patients, who see PGD as "their only chance," are not being sufficiently realistic, there is an equally intense desire to help engendered within the clinical team, who do have something to offer. This, it might be said, makes the issue of "realistic chances" doubly complicated. The team must feel there is a realistic chance for PGD to succeed in order to agree to undertake it. However, they must also ensure that the couple is "being realistic" about the greater likelihood of failure using PGD than any of its alternatives, as well as fully informed about all the things that can go wrong with PGD and their many complicated consequences.

So what does "being realistic" in the context of PGD actually mean? What are "realistic chances," and how can they be calculated? Who, to return to Peter Braude's question, is PGD for? And who is it not for? How will this picture change as more and more people are referred for PGD, and as many of them begin to undertake it toward the beginning of their reproductive life, rather than toward the end?

As discussed further in chapter 4, many of the issues raised in "getting to PGD" are reencountered "going through" it, and indeed as part of "moving on." In continuing to explore some of the distinctive features of "the world of PGD," one learns that many of its key, defining features continue to return in different forms.

Chapter 4
Going Through PGD

Right. They obviously get fertilised on Friday afternoon, here. The couple will get a phone call on the Saturday morning to let them know how many of the eggs have fertilised. Sunday's just a growing day for their embryos. On Monday morning we come in and have a look at the embryos and hopefully assess them for suitability for biopsy. And then they'll be biopsied on Monday afternoon and then tested through the rest of Monday and usually into Tuesday. And then it'll usually be a Tuesday afternoon that we'll do a replacement.

—Jenny Caller, PGD coordinator, Guy's and St. Thomas' Centre for PGD

We got down there, we thought you know, it might not be too bad. But when they explain the options to you, and what sort of percentage chance you've got, you think, "oh that ain't very good is it?"

—Tony, PGD patient

Once a couple have decided to explore the possibility of PGD, and have made an appointment at the clinic, they attend what is known as their initial consultation session to meet with several different members of the PGD team. This session is designed both to provide more detailed information to the couple about what is involved in PGD and to enable the clinical staff to evaluate a couple's suitability for treatment. As Jenny Caller described the purposes of the initial consultation session in an interview: "I suppose in a way I like to say, you know, it's a case of us getting to know you, your story, and what's brought you to us, but also you're getting to know a bit more about PGD and actually what it means, and the fact that it's not a simple treatment, it's hard, and it's not always successful." As Jenny explained further, this session is also designed to ensure that couples are not pursuing PGD as a "magic bullet," or because it was recommended to them by a doctor "desperate to find a cure":

And I don't want to say I spend part of my time trying to talk people out of it, but I—I do think that *out there* PGD is seen maybe as a "magic bullet" sort of thing, if you know what I mean, in the sense that, uhm, there's loads of doctors out there who've been *desperate* to find a cure for this problem of recurrent miscarriage or, you know, wanting to have a child but not wanting to go through, you know, carrying a genetic disorder, wanting to have a child, but not wanting to go through termination, you know, pre-natal [testing] and possible termination. And I think sometimes, PGD, it's just like, "oh, well ok, you'll have PGD." And that's all that's said.

In addition, then, to the need to convey a considerable amount of technical information, the initial consultation session provides an important occasion for couples to "think very seriously about what they're going to put themselves through," "what they might get out of it at the end of the day," and in particular, as Jenny emphasizes, that "actually, you know, their chance of pregnancy is significantly *decreased* through PGD."

Making the Right Choice

For Jenny, then, part of the process of determining whether PGD is the "right" choice for a couple is by *witnessing firsthand* their ability to discuss its potential drawbacks as knowledgeably as its potential promises. Such high standards for critical deliberation of what the PGD option involves cannot simply be measured quantitatively, as an *amount* of information: decision making must be an active, genuine, and serious process, measured by potential patients' abilities to discuss *how they feel* as well as *how much they know*. These *qualitative* evaluations are necessary, Jenny explains, "because some people come here and they say, like I say, they've read everything going, and they've decided PGD is for them: Full stop. And they haven't *thought* about it, you know."

This process of sorting through conflicting feelings about the PGD option connects the past, the present, and the future in a manner that directly shapes the distinctive character of the PGD "choice." What couples have been through prior to their introduction to the PGD option is, as described in the previous chapter, often productive of a unique sense of determination to pursue it as their "only choice." As one PGD patient, Tony, put it succinctly: "Well, I mean we had a choice whether to go for it or not. But in a word, like, there's no other procedure we could go through."

However, as also noted previously, such reasoning may be worrying to members of the PGD clinical staff, who need to feel confident in a couple's awareness that PGD is *not* their "only choice." Moreover, the PGD staff need to have confidence in a couple's ability not only to respond to their present feelings of urgency but to imagine how they are going to feel in the future, especially if PGD fails. As the clinical staff know from experience, it is easy to imagine the relief of being able to begin a pregnancy unaffected by serious genetic disease, but it is far more difficult to envisage the potential costs of failure, the full likelihood of failure, or the possibility that failure might be compounded by the knowledge that PGD was not the "only" choice and that other, forgone choices were more likely to succeed. As Jenny says, unless a couple can envisage these unpleasant possibilities at the outset, there is the risk that "at the other end of treatment cycles, or during treatment cycles, they emotionally can't cope."

Arriving in Clinic

I mean we came out after our first consultation, and it was just *such* a relief. To find somebody who, you know, somebody we thought would really do their best for us, and look after us. (Ian)

Among the couples interviewed, the most prominent and consistent memory of their arrival at the PGD clinic for the first time was one of enormous relief. For many, this arrival represented the culmination of a lengthy search for experts who could provide specialist help. As Ian mentioned above, even though arriving at the clinic is only a beginning, it also feels like an end, in that the couple are finally "going in a particular direction."

An obvious contrast thus arises between the emotional state of arrival—relief—and the purpose of the initial consultation session, which is, in a sense, to plumb this sense of relief for false expectations or unrealistic hopes. Indeed, the initial consultation sessions had been a particularly significant occasion for every couple interviewed, and was one they remembered in minute detail. Several of the interviewees commented on the strong feelings produced by the frank discussions of what could go wrong. Intriguingly, the process of having their hopes severely pruned—which might have been expected to produce a sense of deflation or disappointment—had precisely the reverse effect. It took some time to begin to appreciate why these consultation sessions had the effect of *raising* couples' hopes and expectations of the PGD team, and of the technique itself, and *increasing* patients' confidence in their own ability to cope with its demands.

What began to become clearer over the course of several interviews was that the initial consultation process was where a bond was either established or not between the PGD team and prospective patients—approximately half of whom would go on to undertake PGD and half of whom would not.[1] In an important sense, then, the initial consultation is a *rite of passage*. Even if couples come into the process "certain" that PGD is "right for them," the consultation process itself is designed to test their certainty, in part by emphasizing the many inevitable uncertainties involved in PGD treatment. The consultation process is thus a classic space of liminality, of disorientation betwixt and between a before and after, through which a transformation occurs (Turner 1969). Entering into PGD treatment cannot be a halfhearted default option: it must be carefully considered, and jointly decided upon as a course of action both by patients and by the PGD team after a process that is, in some ways, deliberately unsettling. Having had as their primary feeling upon arrival at the PGD clinic a sense of relief, the very first encounter potential patients have with the team is deliberately designed to "shake them up" and knock their hopes down a peg to get them in shape for what may lie ahead. This process is so sobering that 50% of those who arrive at PGD's door do not return.

Getting a Clear Picture

The transition is equally abrupt for those who stay, who, in a sense, must leave the version of PGD they arrived with and acquire a new one if they are to continue. Moreover, it is only those for whom the postconsultation understanding of PGD is sufficiently convincing who will stay. An obvious question was to ask how and why postconsultation PGD is convincing. A surprise was to discover that prospective patients become convinced, in part, *because of the emphasis on the uncertainty of PGD*.

For Ben and Sally, the feeling of having acquired a "clear picture" of what was involved in PGD and, most important, of "everything that could go wrong" was reassuring at several levels: it relieved them of the false expectation that the PGD team could just "take your eggs, sort your embryos out, and give you children"; it increased their trust in the clinical staff; and it gave them a more complete understanding of the procedure, all of which increased their confidence in their decision to go ahead. The emphasis on the likelihood of the procedure failing, and the detailed descriptions of what could go wrong, made them feel they were

[1] Figures compiled by the Guy's and St. Thomas' PGD team for the period 2000–3 indicate approximately 50% of prospective PGD patients drop out of treatment following the initial consultation session—many simply by no longer turning up for appointments. The reasons for this high rate of dropouts and no-shows are entirely unknown and unexplored, other than anecdotally.

135

being treated with honesty and respect. Understandably, but also some-what paradoxically, *this increased their hopes for success*, as they ex-plained in an interview:

> BEN: And so we had a talk with the Professor and the genetic counsel-lor from the clinic. She went through roughly the same as what our local genetics counsellor had said about what the translocation meant and what not. And he went through what PGD could do for us. They confirmed that they've got the licenses for this specific translocation. And how they do it, this attaching probes to the cer-tain genes and then looking to see this, that and the other, and then he talked us through all the, um, additional tests we have to have afterwards because it's, um, my particular translocation's biggest setback is going to be Down's. So, he said obviously this isn't 100 percent guaranteed that you won't have Down's because it's only a 2 dimensional picture that they take. So, there could be genes lying underneath other genes that they don't see. . . . So, he talked us through all these various, what are they called then—the test for Down's? Amniocentesis, CVS tests and this, that and the other, and the . . . *additional* risk that they bring into it, and would we be will-ing to . . . have the test done?
>
> SALLY: It was a lot clearer that time, wasn't it, to understand?
>
> BEN: Yeah, I mean he spent, he spent a good 2 *hours* with us, didn't he?
>
> SALLY: Yeah, he did. Yeah.
>
> BEN: Which we were really surprised about.
>
> SALLY: [*speaking together*] Making sure we fully understood every-thing before we went ahead. And I think what he was more con-cerned about was . . . I suppose it was like, you think he's just go-ing to take [your eggs], sort your embryos out, and give you children. I think he wanted to make sure that we understood that things could still go wrong. It was very early days and um, it, it was helpful wasn't it? It really was! 'Cause we came out of there feeling quite refreshed actually 'cause we actually both felt we understood. Having spoken to the Professor, and the genetic counsellor, we both understood fully what it meant.[2]

In this account, Sally and Ben's sense of relief is a measure of the transition they underwent from the temptation to think of PGD as a "magic bullet" to a greater awareness of its limitations. Interestingly,

---

[2] Ben's increased understanding of the technique explicitly involved what it *could not do*. As he recollected, "We can't, you can't understand that there's [a limit to what they can test for]—you think, 'oh why can't you do it here?' and 'why can't you do that?' You'd think they're more advanced than they are! I, we did, didn't we?"

and somewhat counterintuitively, this feeling of having had something taken away from them (the scales from their eyes, as it were) seemed like a gain: far from decreasing their hopes for PGD, their more "realistic" appreciation of its limits increased their confidence in its success.

During their initial consultation session, Sally and Ben also gained a fuller understanding of Ben's genetic problem, a more technically detailed understanding of how the genetic diagnosis would be done, and the knowledge that the clinic already had a license for this particular condition (they had done this kind of diagnosis before). They also gained a fuller understanding of how much could go wrong. However, receiving such a large amount of technical information (about two hours' worth), being told that "it's early days yet," and learning how many "things could still go wrong" were experienced by Sally and Ben not as unnerving but as "refreshing." As they both confirmed, their main recollection of coming out of their initial meeting with members of the PGD team was that they "both understood fully what it meant."

"Not a Certainty"

Joanna gives a very similar description of her first appointment and her sense of relief at being told that "the actual chances of it working" were "not a certainty":

> There was Jenny, there was a genetic counsellor, and a nurse, I think. And they just, they talked through everything and showed us—I thought they were, they were *brilliant*. They were very clear, they explained everything really *clearly* and really like, not *patronisingly* but very simply, so you didn't feel silly about asking questions about things. They showed us lots of pictures as well so it was like very easy to understand exactly what was happening. Um, they talked to us about how we felt about it, and sort of, I think, made sure that we understood exactly what was going on and . . . also made it very clear that it wasn't you know, that the, the actual chances of it working . . . are—it's not a certainty. You don't say "Oh I'm having PGD so therefore I'm going to walk away with a baby."

In response to a direct question about why Joanna and Christopher thought that this knowledge was important, Christopher replied by emphasizing the dangers of too much hope:

> I just think that it's so important that you don't build your hopes up, because contrary to what the evening newspapers write about designer babies, it isn't an easy process. It doesn't work in most cases.

And, if you pin all your hopes on it you'll—you know, failure, given what most people who are going through it have already been through, would be devastating for some people.

Christopher's comments convey a cautious attitude toward hoping for too much that is very familiar from other arenas of reproductive assistance, most notably IVF (Franklin 1997). He is self-conscious about the danger of investing too much in the slim hopes offered by PGD. This is a sensible and rational form of emotional protection against the potentially devastating costs of failure—it is a form of control that belongs very much to the complex "ontological choreography" Charis Thompson describes in her account of how patients mold their subjectivities, as well as their bodies and lives, to the demands of treatment (Cussins 1996, 1998; Thompson 2001, 2005). These were actually skills Christopher and Joanna knew well: from their past experience of losing a baby to spinal muscular atrophy (SMA), they had a hard-won respect for the force of the emotionally devastating aftermath of grief and disorientation in the wake of loss. Like Ben and Sally, they found the clinic's honesty about the possibilities of failure reassuring rather than disturbing, in part because its demands asked of them skills they had already acquired through loss. This is a crucial feature of PGD, which, like IVF, is a "hope technology," but it also reveals how both PGD and IVF are conversion technologies, capable of transforming the power of coping with grief and uncertainty in the past into resources for improving the future.

Preparing for the First Cycle

Since it is impossible for anyone, however emotionally self-aware, to imagine going through all the myriad potentially distressing circumstances occasioning even a successful PGD cycle, no one can ever really know for sure how he or she will cope. It is for this reason, as with IVF, that the emotional and psychological demands of PGD are described, by patients and clinical staff alike, as the most challenging, and hardest, part of each cycle. It is for similar reasons that the first cycle is almost always considered to have been the "worst" experience of all.

In addition to the emotional complexity of PGD, its physical, logistical, and technological demands are also considerable, especially for the female partner who must undergo IVF. The usual protocols for IVF, of a period of hormone treatment and ultrasound monitoring, followed by egg retrieval, become more complex with the addition of genetic testing during the in vitro phase of treatment—a period that is extended for PGD. There are thus more chances for things to go wrong in the IVF process for PGD couples—a fact that considerably offsets the "advantage" that, unlike "normal" IVF patients, they are (usually) fertile.

In a standard PGD cycle, the embryos are left to develop for forty-eight hours after fertilization is complete (which may be approximately sixty hours after the sperm are added or injected) before being evaluated for suitability for biopsy. Those embryos with adequate developmental potential are left to continue to divide until they reach the blastocyst stage, when they are biopsied. A single cell removed from each embryo is tested using one of several molecular and cytogenetic diagnostic methods. For single-gene disorders, DNA amplification using PCR (polymerase chain reaction) is undertaken, followed by the application of a probe for one or more known DNA sequences. For chromosomal testing to detect translocations or for sexing, a different technique, flourescence in situ hybridization (FISH) is applied to the nucleus of the biopsied cell, revealing diagnostic features of its chromosomal constitution.[3] The entire process from egg retrieval to embryo transfer normally takes from five to seven days.

However, the "time" of PGD is complicated both by its pace and its intensity. A week might not seem like a long time in itself, but for a couple who may have waited years to undertake PGD, it can seem like an eternity. Similarly, while the "crunch" time of the cycle—from fertilization through transfer or not—is a week or less, this period must be embedded within both its preparation and, in some ways even more important, its aftermath to be fully understood. This before-and-after time is much more amorphous, especially as cycles accumulate and blend together into a sea of unfulfilled aspirations.[4]

The time of PGD, like the time of IVF, is also difficult to measure in strictly chronological terms owing to its intensity. Mentally and emotionally, the experience of PGD is consistently described as "preoccupying"—both because of the high levels of investment in its outcome and because of the sheer amount of labor, physical and otherwise, involved in a cycle:

TONY: It's, I don't now. You see, it's hard work innit Mel?
MELISSA: [*in the kitchen*] What's that?
TONY: It just seems to play on your mind a bit. [*raises voice so Melissa can hear*] It just takes over your life.
MELISSA: Yeah.

---

[3] FISH is a cytogenetic technique that uses a probe tagged with a fluorescent dye to bind to targeted regions of particular chromosomes (See Scriven et al. 1998).

[4] A "cycle" is also difficult to define. In addition to the time required for the IVF cycle to stimulate and collect eggs, a significant period of preparation, mentally and physically, for a cycle is typical. Thus, although a cycle takes less than a month, it can dominate the better part of a year.

Tony and Melissa's description of PGD as "hard work," as "play[ing] on your mind," and as "tak[ing] over your life" are typical of the descriptions of going through a PGD cycle, in which the "unbearable wait" of IVF is extended even further.[5]

The Cost of a Cycle

Preparation for a cycle includes raising the funds to pay for it—a source of delay to any couple unable to meet the four-to-five-thousand-pound cost of treatment as well as the added expenses of time off work, travel, and accommodation, which for couples living outside London could double the cost per cycle. The financial commitment required by PGD also creates a difficult dilemma for some couples who worry about spending all their money getting pregnant and then being in long-term debt or not having enough to care for potential children. Sally—who was in her late twenties when interviewed—described this concern:

> SALLY: I was getting worried. . . . I thought to myself, I thought if we've got to pay for the IVF and if we are very lucky, and we do get a pregnancy, what if the PGD doesn't pick up and we get a Down's syndrome baby? I was thinking in the back of my mind, what would happen if we have to bring up a Down's syndrome baby and we've got to provide its care for its, you know, future in life? I was thinking "how are we going to have all this money?" Because all this money we're spending on IVF in the first place! And that's what my main worry was, and I was getting a bit stressed out about the whole lot really!
>
> SF: That it would be an ironic position to be in that you've spent so much to have a child and then the child really needs the money spent on it.
>
> SALLY: Yeah, you can see what I'm saying, yeah. So, when we got funding [for a second cycle] it did take the pressure off because I was thinking that okay, we've saved about four and a half thousand pound, wasn't it, with the drugs? We've saved that amount of money. We haven't *saved* it because we haven't *got* it, but we haven't had to fork out that amount of money, so that money that

---

[5] This form of conversation between Tony and Melissa, in which the two speak each other's experiences, often complete each other's sentences, and conversationally produce a bond as "one voice" was a striking and repeated feature of the interview set—made particularly vivid in its transcribed form, and itself indicative of the bond PGD produces even when it fails. This form of conjugal bonding through reproductive labor and the creation of "partial life" is also evident in couples' descriptions of their embryos, as is discussed further below (and see Thompson 2005).

we would be saving every month towards IVF can go aside into an account. Because if something like that *did* happen or let's just say in the future we wanted another IVF attempt that you know, we could pay for. It's the worry of getting in debt. I was thinking of getting into debt and having children, I know it's a lot of people are in debt aren't they, with children, but we don't want to be in a great amount of debt.

Funding concerns comprise an important set of issues for patients, and are often the source of delays in initiating or continuing treatment. When a couple decide to go ahead with a PGD cycle, a request is sent to the patients' local NHS Trust for funding. Over the course of this study, these requests were 50% successful for one or more cycles, although in general fewer than a third of PGD patients in the United Kingdom are estimated to be publicly funded. Couples whose requests for NHS funding are declined have no option except to pay for treatment themselves, often with help from their parents. Some couples who did not receive funding for cycles did receive free or subsidized prescriptions for ovarian stimulating drugs from their general practitioner. These drugs cost about nine hundred pounds.[6]

"It's Hard to Understand if You Aren't Sniffing"

Once funding issues have been resolved, treatment can begin. The PGD cycle begins with women taking specific medications to "down-regulate" ovarian production of hormones—the standard protocol for IVF. These agents produce a temporary chemical "menopause" with its accompanying symptoms of mood changes, hot flashes, sleeplessness, and irritability. Some couples, such as Tony and Melissa, described these symptoms of "the IVF side" of PGD humorously.

> SF: So, that first cycle, did you find that you could cope with the IVF ok, or was it more complicated than you thought?
> MELISSA: I could cope with the IVF side of it. It's just the sniffing drugs. I can't cope with all the sniffing drugs! [*laughter*]
> SF: Yeah, right.
> MELISSA: I'm an absolute nightmare [*laughter*] I am! I *really*, really, I just go into this—

[6] For some couples, the financial costs associated with PGD were significant, even if they had received funding from their local health authority. Traveling to London, staying in a hotel, and taking time off work were all expensive, and some couples traveled to London from as far away as Scotland. For these couples, as is discussed in chapter 5, financial costs played an important role in deciding when to end treatment.

TONY: [*interrupts*] I spend twenty hours at work!

MELISSA: Yeah, he does! [*laughter and joking*]

TONY: [*interrupts*] I have to hide all the knives and everything! [*laughter*]

CR: You're looking forward to menopause then? [*laughter*]

TONY: No! [*laughter*].

Mood changes caused by the hormonal component of IVF treatment were often a source of joking between partners; they also frequently featured prominently in interviews, and their effects were prominent in descriptions of PGD.[7] For some women their "emotional" state had led to difficulties at work, while others had to stop working altogether. Melissa's emotional volatility had led to her taking a leave of absence from work because it was affecting her "quite badly."

TONY: I mean Melissa's really got a nice personality but I, I can't believe how you change! Evil or—!

MELISSA: After about four or five days of it. But the thing is I know myself that I'm doing [it], and I just couldn't care less! But that's just—it's just hormones! I mean I can go black in one minute and—I mean, last time . . . 'cause they'd kept me on it longer as well,[8] it was really affecting me at work quite badly. . . .

TONY: I think it was *four* weeks they left you sniffing for because they couldn't get us in [for an egg retrieval appointment]. And they said "Right just sniff for another couple of weeks." Oh great! [*smiling*] And I come home and you says . . . "Jenny's told me that I've got to sniff for an extra couple of weeks." Cor blimey! Thank you!

MELISSA: So it got quite bad at work. I, I did have to go home—

TONY: You're more upset, aren't you [when you're sniffing]?

MELISSA: Yeah.

TONY: Happy one minute and then puh! But it's hard to understand, if you aren't sniffing.

The mood changes and work disruption caused by the hormonal treatments required for IVF may be compounded by the amounts of time required to travel to and from clinic for ultrasound scans to monitor

[7] A number of couples expressed concern about the potentially carcinogenic effects of the drugs used in IVF, and a desire to limit their PGD attempts to minimize this risk, as is explored further in chapter 5.

[8] The "down-regulating" period of IVF to suppress the normal menstrual cycle is followed by the introduction of hormones to stimulate follicular growth, initiating a sequence of events that must be precisely timed to coincide with the work schedules not only of embryologists and clinicians but the geneticists who will need to conduct their assays within a narrow twenty-four-hour window. This is why the scheduling of PGD cycles sometimes requires longer periods of "sniffing."

follicular growth, which is simultaneously being stimulated by a series of hormonal injections to induce ovulation. For PGD, even more than "ordinary" IVF, it is imperative to maximize egg production in order to generate the largest possible number of embryos. However, it is also essential to avoid overstimulating the ovaries, which can cause ovarian hyperstimulation syndrome (OHSS)—a serious condition requiring hospitalization, and almost certainly necessitating that the cycle be abandoned. This "fine line" of IVF treatment, although increasingly subject to more sophisticated management and prevention protocols, is a particular concern in the context of PGD, in which maximizing egg production is at a premium.

"It Doesn't Make Any Difference Whether You're Fertile or Not"

The outcomes of hormonal stimulation and egg retrieval are both unpredictable and varied, and often an initial failed cycle will yield valuable insights into the correct hormone dosage in subsequent cycles. Some women reported being surprised at their responses to these drugs, producing up to twenty-five eggs. Others inexplicably had no response at all, which led to the disappointment of the cycle being abandoned.

For many couples, the fact that IVF success rates were not significantly higher for the largely fertile cohort of PGD patients was an unpleasant discovery, removing one of the few perceived "advantages" of PGD over "normal IVF." While implantation rates of embryos appear to be higher for couples without a history of infertility, this is an area of uncertainty within PGD for both couples and clinicians. As Jenny Caller explained, this is because IVF, ironically, induces the very infertility in fertile patients that it is designed to alleviate in infertile ones:

> One of the really fascinating things about my job is that we are doing
> IVF work on people who are fertile. And you'd think "ah, there
> won't be any problem getting eggs, there won't be any problems, you
> know, fertilising eggs. There won't be any problem getting embryos."
> But it doesn't work like that! Because what we're doing, we're taking
> everything out of the body and manipulating it, and using the drugs
> in the way that we do is so unnatural it doesn't make any difference
> whether you're fertile or not.

This observation, that IVF does not appear to be significantly more effective for fertile couples, raises a number of questions about the technique and its relationship to in/fertility. For Jenny there are several possibilities. At one level, IVF is "asking women's ovaries to do something completely unnatural," but at another level humans are "not a fertile species, nowhere near." Indeed, if the chance of becoming pregnant for the average couple having unprotected intercourse is 20% in any given month, "maybe IVF is putting them back on the same level, it's just that human fertility is not that good."[9]

These questions are shared by patients, who frequently wondered about, and often queried, aspects of IVF treatment, such as the hormone-level dosages and timing. This is often an unexpected source of concern for PGD couples, most of whom have a history of proven fertility. As Jenny Caller explains, "every single couple who comes through this clinic thinks that because they are fertile it will be easier":

> They don't see that, you know, it's using the drugs to get follicles and to get eggs, to get embryos, that's going to be a problem. What they *worry* about, and focus on, is the *testing* of those embryos, and how many are going to be unaffected. And that really, to me, gives you a good indication of their perceptions of their risk. Because they are really worried that there won't be embryos suitable for replacement. So what they've got in their head is "every time I get pregnant the baby's going to be affected."

Jenny's comments about risk perception in the context of PGD clearly articulate two important themes that emerged in a more amorphous and indirect manner from the interviews. The first is that while the focus at the outset is on how difficult PGD will be, it is actually IVF that is the cause of much treatment failure—and for reasons that are not fully understood. The second theme, which is a consequence of the first, is that while PGD is offered to alleviate a condition of *genetic risk*, it may result in an increased awareness of *reproductive failure*. This can result either from the discovery that a couple's genetic condition directly affects their fertility (as is the case for some male CF carriers) or from the experience of unexplained reproductive failure in the context of IVF.

Wait Control

Egg collection sets off a tightly choreographed chain of events, punctuated by bursts of frenetic activity that mark the key stages of PGD. These are separated by oceans of waiting, about which many patients had vivid memories.

[9] The substitutability between the "assisted fertility" of infertile couples treated with IVF, the natural "infertility" of humans in general, and the ambiguous fertility of fertile, semifertile, and infertile couples undergoing IVF for PGD is, interestingly, reflected in the approximately 20% success rate that characterizes all of them.

The first important marker, and the baseline of any given PGD cycle, is the number of eggs retrieved at the time of collection, which for the couples in this study varied from none to twenty-five (more often it is in the vicinity of ten to twelve). At the time of egg collection, the male partner is asked to provide semen, from which sperm are extracted, cleaned, and injected into the oocyte using intracytoplasmic spermatic injection (ICSI)—a micromanipulation technique normally used as a form of male infertility treatment, but which in PGD is necessary to avoid contaminating the outside of the egg with paternal DNA that might compromise the accuracy of the genetic assay. ICSI is followed by "in vitro fertilization"— a process that takes approximately twenty-four hours, during which the DNA of the egg and sperm are repeatedly split and recombined.[10] The second piece of vital information couples receive is how many of the eggs have fertilized and how they "look" to the embryologists. "Good-looking" embryos display clear, even, well-rounded development, giving a robust appearance of vitality. "Poor-looking" embryos show signs of disaggregation, uneven development, opaque coloring, and stalled or delayed cellular division.

For PGD, embryos must not only appear to be developing successfully but must continue to do so until they are large enough to withstand the removal of a single cell for biopsy. The third vital sign, then, will be how many of the embryos that have successfully fertilized and developed are suitable for biopsy. Following biopsy—a technically demanding and extremely laborious process, particularly with large numbers of embryos—another degree of loss will almost inevitably be registered, as not all the embryos survive biopsy. Some of them will "lyse" or burst: others will collapse, fall out of their shells, or behave unpredictably. For other reasons, not all the embryos are biopsiable—for example if they are too "sticky" or have begun to develop the "tight junctions" that precede compaction in the blastocyst phase.[11]

This technically complex and intensive period of treatment is the crucial buildup to the whole "point" of PGD—and the moment for which it is named—which is the genetic diagnosis of the preimplantation embryo. This is the most agonizing period for patients, who, having been partners with clinicians in the effort so far, are relegated to the sidelines, or more accurately the phone lines, to wait—unable to contribute anything but their restraint and self-control. While the clinicians, embryologists,

[10] Embryos that will be screened using PCR must be fertilized with ICSI to prevent contamination, although those that will be sexed or tested using FISH can be left to fertilize "normally" in vitro.

[11] The need for the embryos to be as large as possible in order to survive biopsy is in tension with the need to biopsy early enough to avoid the "tight junctions" of later development, which prevent the removal of a single cell.

and geneticists are now entering the most active, dramatic, and decisive part of PGD, the patients are relegated to an unpleasant state of passivity. This state of "doing nothing" contrasts sharply with the sense of momentum and of being "picked up by the system," which was frequently described by patients as one of the most rewarding aspects of treatment, as in Ian's account of how good it felt to be moving forward "in a particular direction" after years of reproductive difficulties:

> Once you started being picked up by the system, it developed a momentum all of its own anyway, so there's this sort of constant reassurance we were getting. You know about—I sort of keep re-emphasising the fact that once we were involved with the people at St Thomas', they were very positive and very encouraging really, and patient really, give you time to ask them questions, so I felt that once you—you seemed to come on board, really, and you were carried by a lot of it. And it—in a way it's kind of a relief, to have that, to have someone sort of directing you—you're going in a particular direction.

As Ian explains, this sense of momentum and direction is created through personal interaction with the PGD team. Their encouragement and reassurance, as well as their patience and responsiveness, create a sense of being carried along, being taken on board, and of being "involved." Part of the challenge of going through PGD, then, is the need to adjust to an abrupt shift toward the end of the treatment cycle from being at the center of the action to a feeling of being left behind. Meanwhile, it is during the post–egg collection period that patients' friends and families will, like them, be on tenterhooks. Will it work? Will all the effort pay off? Will the odds be in their favor? Will the gods be on their side? The phone, at once a treasured lifeline to the clinic and to any news of progress, is also a dreaded source of potential disappointment and a constant source of well-meaning, but difficult to manage, inquiries as to "how things are going." It is, unfortunately, impossible to tell, as even the best odds can deteriorate rapidly, and the worst can produce the best results. In this respect, for patients as well as their anxious families and friends, there is nothing to do but wait.

"Have They Phoned Yet?"

For all these reasons, patients inevitably described this part of the PGD cycle as agonizing. Significantly, this was also a period in which the PGD couples' attachment to their embryos was a constant preoccupation. Instead of visiting the clinic themselves, for routine scans and consultations, the patients had, in a sense, been replaced by their embryos, who were now the object of the PGD team's attention. Having left a part of themselves behind in clinic, Tony and Melissa drew a complicated picture of phone lines connecting them both to overeager friends and to the clinicians tending their precious embryos, toward whom they did not want to appear overeager themselves.

TONY: I *cannot* breathe, I cannot *nothing*! Every time that phone rings and you pick it up—"Hello! Have they phoned yet?" [*imitates family member*] And you go "Oh god!" [*highly exasperated tone*] [*laughter*]

MELISSA: Get off the phone!! [*laughter and joking*] . . . Oh dear me! People drive you mad, don't they? They're all just waiting. Nobody will go out anywhere! Everybody's just stuck in!

TONY: And the phone, [it rings] all the time and you go, "Oh god!" 'Cause your heart's in your mouth isn't it. And they go, "Have they phoned?"

MELISSA: And then you get to the stage where, if they said yeah, that you've got some clear, get back here [*Tony sighs*] at such and such a time, we book the train . . . all the while you're going there you think to yourself "Well, overnight, have, have they dropped out? Or has something happened?"

TONY: It's terrible! I mean, really you want a twenty four hour line to them saying, "How they doing now?" "How they doing now?" You must get on their nerves—I bet they're saying "Leave us alone!" [*laughter*]

In this extract, which is typical of many accounts of the "terrible" experience of waiting by the phone, Tony describes his anxiety level as being so high he "cannot breathe" and indeed "cannot nothing." His and Melissa's painful sense of anxious waiting is exacerbated by phone calls from their family and friends who are doing the same thing and have nothing else to do but phone for updates. In what almost resembles a farcical relay, the phone is picked up by Tony, his heart in his mouth expectant for news, only to be asked "have they phoned?" by a family member. Needing both to maintain their highest hopes and to protect themselves against the very same, they must book their train in case they need to rush down to London for the embryo transfer (the best-case scenario), all the while wondering "have they dropped out?" or "has something happened?"—which is the dreaded, but statistically most likely, outcome.

Tony and Melissa's almost umbilical image of a twenty-four-hour line to their embryos is not surprising, given that they are physically so far away from clinic just at the penultimate point when the emotional denouement of the cycle is about to begin. Desperate for a connection to their embryos they fear becoming, like their well-meaning friends-who-phone-too-often, a potential source of unwanted distraction to the PGD team, imagining, "You must get on their nerves—I bet they're saying 'Leave us alone!' "

Since, as happens in all the other stages and procedures in PGD following egg collection, both PCR and FISH will further reduce the number of viable embryos, and may eliminate any hope of transfer altogether, another feature of the waiting-by-the-phone period that is difficult to bear is the inevitability of progressive losses, or "dropping out," from the precious and limited embryo supply. In other words, the couple must leave the clinic when their gamete supply is at its highest, wait while it is depleted by the inevitable casualties of treatment, and hope there will be enough "clear" and surviving embryos for the PGD team to collect.

All these factors explain why being separated from their embryos is difficult for many couples to deal with. As Melissa adds, what they really would like to do is to "sit there all night," watching the embryos growing and dividing.

> MELISSA: I mean the first time that . . . we got some sperm and that they'd gone together and we knew we'd got a few to test on, I felt like I wanted to *sit* there all night—
> TONY: With them
> SF: And be near them.
> MELISSA:—and all day, every day and just watch them. And that's how you feel and you've just got to think no, just come away.

This sense of wanting to be with their embryos is also a desire to be physically present in clinic, much as people describe wanting to remain in the hospital during medical procedures on friends or relatives. The desire for physical proximity, in contrast to being miles away, is both an expression of hope and a yearning for agency.

As Joanna explains, "someone else is doing the stuff and you're just waiting by the telephone":

> Once those eggs are taken out of your body, *it's all happening somewhere else*! So, at least when, when you're taking the drugs, even though it is out of your control in a *way*, because it's happening inside *you* and *you're* the one injecting the drugs, and you're the one to be deciding—even things like what you eat and whether you're drinking or not drinking or how fit you are—all those kind of things, you're feeling like you have some *level* of control—but once it's out of your body that's it—it's someone else, somewhere else, is *doing the stuff* and you're just sat there waiting by the telephone.

Other couples also described this time as "the worst part of PGD," because it was at this point that the outcome felt most "in the lap of the gods." As Natalie recalled, this was the period during which she felt things were most out of her control:

> Once they've taken the eggs from me and they do the research, it is really in the lap of the gods, isn't it, as to whether or not they have one or two, let's hope three [laugh] . . . , to go back. . . . That is *totally* out of our control . . . there's nothing we can do once, once that's done.

For Joanna, who was undergoing PGD after having lost a daughter, Sasha, to genetic disease, the wait for the phone call from the clinic reporting on the results of the genetic tests on her embryos unavoidably recalled an earlier period of life-and-death results concerning Sasha's diagnosis. As she recalls:

> Well that was on a *Saturday*, so then we had to come back [from the clinic] and we just like spent the Sunday like this [sounds depressed]. Monday *afternoon* they rang. Monday was *horrible*! 'Cause we waited for phone calls and I think that's the worst bit is just sitting waiting for the phone to ring. And it, it reminded me actually of when we waited—when we were waiting for the doctor to ring with with Sasha's results. We knew he was going to ring on a particular day, and you just sit there just waiting for this phone call, and you *can't* concentrate on anything. And you just . . . feeling sick and, and I'm *terrible* anyway because I tend to—I can't—I tend to analyse *everything* so I'm sort of my heads going "this could happen, that could happen and if this happens that happens." [*imitates anxious tone—little laugh*]

As well as the life-and-death stakes of PGD in the present that were so preoccupying for Joanna, so were its connections to a painful past, and in particular to the death of her daughter Sasha, by which the future genetic diagnosis of any surviving PGD embryos would be marked, by either repetition of or escape from that possibility. This complicated temporality of PGD—the way it fuses losses in the past with preoccupations in the present and potential gains in the future—is one of the most vivid experiential features of PGD, through which even time and space acquire new dimensions.

Once either cytogenetic or molecular testing is completed, couples are called back to the clinic to discuss the results with Professor Braude and one of the two geneticists on the PGD team who conducted the diagnostic tests (Paul Scriven or Caroline Ogilvie). It is at this point that a decision must be reached about which embryos to transfer, if any. This is a decision in which the couples' role is crucial, whereas the technical complexities of the procedures, and their interpretation, require considerable expert knowledge. It is thus, at this very emotional and critically important penultimate moment, that PGD couples are reunited with the top clinical staff from both "sides" of PGD—reproductive and genetic—to make what is literally a life-or-death decision.

The Genetic Diagnosis

As described in chapter 1, the genetic diagnosis required for PGD relies on vanishingly minute amounts of DNA, as well as the premise that each single cell is genetically identical to the rest of the embryo. Unlike prenatal diagnosis (PND), such as amniocentesis or chorionic villus sampling, "in which relatively large quantities of pure genomic DNA can be extracted from biopsied tissue samples, which are made up of many thousands of cells," enabling not only "direct diagnosis" of genetic or chromosomal material but also repeated testing "several times over a period of several days" to confirm the results (Braude et al. 2002), PGD depends on the contents of one, or at most two, cells.

In order to optimize the results of genetic analysis on such a minute quantity of DNA, it is essential that stringent precautions be followed to avoid contamination by extraneous, nonembryonic DNA, including isolation of PGD laboratories and equipment, air filtration, and the use of a gown, mask, and gloves by laboratory staff. Such precautions are among the many exacting requirements of maintaining a successful PGD program and minimizing the dreaded-but-inevitable incidence of misdiagnosis. Even with these rigorous procedures, however, the potential for misdiagnosis remains intrinsic to features of the testing process itself (which is continually evolving as new refinements are made). For example, the amplification of target sequences may inexplicably fail, causing a heterozygous locus to appear homozygous—an error that could lead to either the false elimination of an unaffected embryo or the implantation of an affected one.[12] Patients must consequently acquire a working knowledge *both of how the testing techniques are supposed work and of how they may fail*. This knowledge becomes especially important at the point of decision making about which embryos to transfer, because the decisions are often not entirely straightforward.

[12] This artifact of PCR amplification, known as allelic dropout (ADO), requires that additional means of identifying target sequences and attesting to the provenance of samples have been developed, such as multiplex PCR, which "affords the opportunity to develop a generic strategy for a particular disease, which is independent of the mutation present" (Braude et al. 2002, 946).

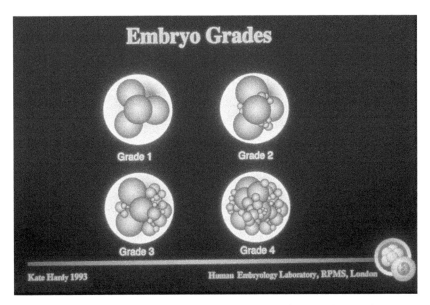

FIGURE 4.1. In ordinary IVF, fertilized eggs are "graded" according to their morphological appearance in order to determine which are the best eggs to be transferred to the womb.

To explain and illustrate the eagerly awaited results of the embryo testing, and to make the crucial decision of which embryos to select for transfer, the clinicians and scientists on the PGD team use several different kinds of images of embryos. Each embryo is numbered, and the test results are shown as a sequence.

FISH results are in color, showing the bright stars of fluorescence that provide the basis for determining the genetic or chromosomal constitution of each embryo. In addition, the team uses black-and-white images of the embryos to assess their morphology. As well as being classified in relation to the genetic condition as "clear," "carrier," "affected," or "ambiguous," the embryos are graded according to morphologic criteria indicating their likely potential for further development after transfer.

A classically difficult PGD decision arises when the "best" embryos morphologically are not the most desirable genetically, or vice versa. During Tony and Melissa's third PGD cycle, which unfortunately ended in miscarriage, seven embryos suitable for biopsy were created from twenty eggs. Four of the embryos were suitable for transfer, but the two that were carriers of the cystic fibrosis mutation were superior morphologically to the two that were "clear," or unaffected by the single-gene mutation. Although many people can carry the CF mutation and never suffer health consequences, such as Melissa, other heterozygous CF carriers, such as Tony, experience significant adverse effects (in his case subfertility). The dilemma, then, concerned which embryos to select for transfer—those that were morphologically better but genetically affected, or the unaffected embryos that were "clear" but did not look quite as "good"? This was a decision in which the PGD team and the patients were equally involved, as Melissa and Tony explain:

> MELISSA: 'Cause last time we'd actually got *four* good embryos hadn't we and . . . we were called back and we were told that we'd got two clear, but before we actually went and had the embryos put back um,
>
> TONY: [*interrupts*] The *worst* thing was—they'd got them hadn't they? They showed the photos of the eggs and they'd got four. [Professor Braude] said "It is the worst scenario for us *ever*." He said "because we've got two clear eggs here" and he said "we've got two carriers." But we don't know the extent—how they're carrying. It could be like me, it could be like Melissa. Now if they're like Melissa, no problem. The two *best* eggs were the two carriers.
>
> MELISSA: And they *looked brilliant*.
>
> TONY: Absolutely fantastic.
>
> MELISSA: You could *really* see the difference.
>
> TONY: The other two, weren't as good as them two.
>
> MELISSA: They were more granular weren't they? And like, Professor Braude says, you know, "If, if we hadn't done this test we would put the two back that were the carriers."
>
> TONY: In fact, they were gonna, they were struggling with us at one stage weren't they? *That* bad. Professor Braude says "We can't tell now by looking at the eggs whether it's got full blown cystic fibrosis or not." And we had to decide.
>
> CR: Yeah. What did you do?
>
> TONY: The two clear, didn't we?
>
> MELISSA: Put the two clear back. And Professor Braude was *really* pleased, he says to me "We've come this far, we're glad that you've chosen to put the two clear back."

Tony and Melissa's decision to select the genetically unaffected, but morphologically uncertain, embryos is consistent with the overall logic of PGD—as is affirmed by the clinic staff. However, the conflict between the reproductive and genetic "sides" of PGD is clearly demonstrated by their dilemma, to which there is no uncomplicated resolution.

Recognizing the importance of patients' participation in decision making, and making the commitment to their fullest participation possible, is one of the most difficult aspects of PGD to achieve. Clinical best practice in situations such as these requires a tightly integrated team of highly trained specialists who also have excellent communication skills, and who are willing to give their time not only to the demanding protocols of PGD but to explaining these to patients. Since clinical staff are always short of time, frequently behind schedule, and invariably working under unpredictable circumstances, the commitment to involving patients as fully as possible in all aspects of PGD decision making has to be a genuinely shared priority to become a reality.

When PGD is successful, no amount of time or effort, expense or discomfort, could possibly seem in retrospect to have been "wasted." However, it is because PGD so often fails that *the same is also true when PGD does not succeed.* In Tony and Melissa's case, their active participation in decision making, their trust in the PGD staff, and their appreciation of the honesty and unstinting explicitness of their conversations with members of the PGD team throughout their treatment enabled them to feel confident the right decisions had been made, even though their treatment failed. Their ability to recall the exact appearance of their embryos at the time they made their decision, and to explain in detail the technical basis of their decision in retrospect, gives the impression that these details have proven helpful in their coming to terms with failure, which in this case has involved literally framing the portraits of their "best" embryos as a reminder of their partial success.

Would You Like to See Our Embryos?

Tony and Melissa's account of their difficult decision at the time of embryo transfer was typical of many of patient interviewees in its emphasis on the importance of seeing pictures of their embryos, and typical also in its testimony to the lasting significance of their attachment to these images, formed at this critical juncture in their lengthy journey to achieve a successful reproductive outcome. Many couples had kept their embryo pictures, and frequently took them out during interviews—if they were not already proudly displayed on the mantelpiece. The role these images played in couples' lives was obviously both profound and complex. At once records of achievement in having successfully passed through all the stages of IVF, they were also visual documents of almost-offspring, of partially successful procreation, and thus of a conjugal bond made of love, labor, the creation of new life, and its loss. Keepsakes from their time in clinic, when the photographs would have been taken, and used in the decision of which to transfer, the embryo images had "come home" with their parents, to live with them and share their PGD histories. Like the time and space of PGD, these images were dense nodes of collision between the historical and future present, achieved and "normal" conception, public and domestic identities, and a sense of pride mixed with loss.

The presentation of these images also "performed" these mixed identities. Some embryo pictures were presented in much the same way clinicians and scientists would do—to illustrate the difference between an affected and an unaffected embryo, for example, or to explain a particular translocation. At other times, the embryos acquired more complicated identities. A genetically affected embryo might stand for the disease itself, or a child who had died, or the fear that had prevented a couple from trying to conceive another pregnancy. In these cases, it was not so much a *picture* of an embryo but *the embryo itself*, and the life or death it symbolized, that was being presented.

Trish and Ian's embryos were so "picture perfect" that they were referred to by the clinic staff as "textbook embryos." In the following extract, her comments about her "textbook" embryo pictures are characteristically multilayered. In her narrative laced with pride as well as grief, and describing her embryos as "lovely," Trish relates her embryo countdown, from fifteen to two to none.

> I've actually got the pictures! [*laugh*] . . . We don't have pictures of family up here, we just have our "textbook" embryos! [*laughter*] Those were the fertilised embryos. [*showing pictures*] There were 25 eggs, 13, 15 took fertilisation but two of them were a bit dodgy, so those are the 13. That's the fluorescence [*in situ hybridization image*]. . . . It was great because we were just given all this information. And these are the two that they transferred. [*showing pictures*] I mean to me they all just, especially the lovely coloured pictures, [*laugh*] you know, they just look *lovely*! Um, and 8, yeah, 8 embryos were left to choose from, which were either normal or carrier.

Here, the blurred boundary between the (retained) image of the embryo and the (lost) life it represents is dense with contradictory implications, as in Trish's ironic, but painful, aside: "We don't have pictures of family up here, we just have our 'textbook' embryos!" This comment responds to the characterization of the embryos as so "perfect" they could be textbook illustrations, which Trish and Ian understandably take as a compliment.

The implicit tragedy of the pictures on Trish's mantelpiece is offset by

---

[13] For a discussion of the significance of dead embryos in the context of the work of British artist Helen Chadwick, see Franklin 1999a and see a discussion of embryo disposal in Franklin 1999b.

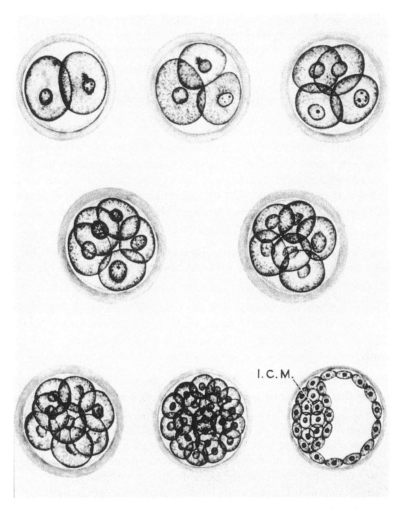

FIGURE 4.2. These hand-drawn images of embryos from an embryology text-book represent the "ideal" form of embryonic development.

her ironic delivery and her emphasis on the beauty of her embryos, which is a lasting source of pride in their existence, their perfection, and their loveliness, while also being a memento of their loss.[13] In admiring their loveliness ("they just look lovely"), Trish refers to both their acclaimed 'textbook" perfection and their "lovely" colors.

The beauty of these embryos is both a natural and a technical achievement. They are naturally so beautifully formed that they could be used to exemplify perfection, and they look even lovelier still because they have been given "lovely" colors. As objects, the photographs are lovely both for what they represent (their content) and for their technical precision (their form). They affirm a unique accomplishment, while they are also painful reminders of loss, of failed potential, and of grief. They are, moreover, intensely personal and intimate images, while also being artifacts of a highly technical clinical procedure conducted in a sterile laboratory.

These complicated dimensions of patients' attachments to their embryo pictures thus confirm many of the distinctive features of PGD—its before, during, and after—as well as many of the technique's most contradictory or paradoxical aspects. For Joanna, the experience of seeing an affected embryo was both "weird" and disconcerting. Describing the impact of looking at the PCR test results—the large blank gap in the electrophoresis gel "bar code" indicating an affected embryo—Joanna recalls:

> And they showed us the, sort of bar code thing, you know [smiling]. And um, it was really *clear*, it's really weird 'cause it was really clear, there was just this *gap*!

She subsequently describes seeing the enlarged images of her embryos as "traumatic" and explains, along with her husband Christopher, how seeing the affected embryos disturbed her psychologically and emotionally:

> JOANNA: Well, it was the one that was *affected* that actually affected *me* the most because it was like—sort of knowing that—that that was a potential SMA baby. And that was like ooh [*worried tone, little laugh*]. And it, and it was looking at these sort of living— pictures of living cells—all of a sudden it just hit me that these were actually potential *babies*. And I think psychologically I hadn't thought about it like that before. It sort of— [*pause*]
>
> CHRISTOPHER: I, I'd thought about it, but when you actually *see* this and then you get all the *questions* around IVF and termination, you know, be it at the 8 cell stage or . . . at 3 months, you know, it is a um, potential um, potential life. And you, you've got to be aware, I mean well having *lost* a baby [*little laugh*] you are very much *aware* of it. Um, but you've also already been through a—losing a baby through a genetic condition which um, *it is* very, very draining, very *traumatic* when you know that the baby's going to die. It brings it home to whether *you* could go through that again. But just looking at those little cells, it's almost a case of, if that takes off, well they're our babies.

For Joanna and Christopher, the impact of being able to see which embryos were affected—both on the PCR gel and in the picture of the affected embryo—threw them back in time to the birth and subsequent death of their baby daughter, Sasha, from spinal muscular atrophy. For both Joanna and Christopher it was the potential of the affected embryo to become a terminally ill child that could be "seen" via the process of PGD, in the same way that waiting by the phone for her PGD results reminded Joanna of waiting for the doctor to call about Sasha's genetic diagnosis many years earlier. "Just looking at those little cells," Christopher explained, reminded him of what it felt like "when you know that the baby's going to die" and brought "home" once again the trauma of that experience and the question of "whether you could go through that again." As he continues:

> Yeah. It was, it was pretty much a—that's a, you know, a little basically a little Sasha, you know, well surely that's how *she* started out. So, there are—I mean we, we *said* one of the reasons we went to PGD was if we were to have the CVS test, that at 3 months we'd have to make a decision to abort a *beautiful* little child! That we know is going to have a limited life. . . . [With] PGD, you're taking that decision 3 months *forward*, before you've got to the stage of having that child developing inside you and *knowing*, um, that you've got to make the decision to abort or go ahead with a child who's going to live for 9 to 18 months! And die in a not a particularly pleasant way. You know. [pause]

For Christopher, the experience of looking at an affected embryo generated vividly emotional moral questions about its simultaneously "beautiful" and lethal potential. Because the affected embryo was "basically a little Sasha" and was "surely how she started out," the affected embryo was at once "a beautiful little child" and a child who was going to live only for a matter of months and who would "die in a not particularly pleasant way." For Christopher and Joanna both, it was the ability to have the knowledge that an embryo was affected "before you've got to the stage of having that child developing inside you" that was the crucial difference PGD made. Seeing an affected embryo, and the "clarity" of the PCR gel through which it was diagnosed, simultaneously confirmed for them the value of PGD and reminded them of the tragic circumstances that had brought them to its door. They were faced here simultaneously with the reason they were doing PGD—to avoid giving birth to another affected child—*and* with the logical conclusion that even PGD involves the necessity of selection, termination, and death. Although the death here—of an eight-cell embryo—was both morally and emotionally preferable to terminating a three-month-old foetus, it was nonetheless upsetting, as it was inevitably linked with Sasha's "beautiful" life as well as her tragically premature and painful death.

PGD Pregnancies

The difficult life-and-death choices involved in the preimplantation stage of PGD—which led Christopher and Joanna to its pursuit in the first place—are encountered in a different way if a pregnancy is established and a choice must be made about prenatal diagnosis (PND). If a pregnancy is achieved, the Guy's and St. Thomas' clinic offers PGD couples the choice of having a chorionic villus sampling (CVS) or amniocentesis test to ensure that the genetic diagnosis was accurate. The clinic, although technically neutral on the matter, favors PND as a final confirmation and because, like Tony and Melissa's choice to have the unaffected embryos transferred, it is consistent with the overall logic of treatment. However, for any couple who has gone through the considerable effort of PGD and been fortunate enough to achieve a pregnancy, the thought of endangering that pregnancy in any way is unattractive, to say the least. For many patients, moreover, the decision to have PGD was precisely motivated by their desire to avoid PND, the "tentative" pregnancy it requires, and the haunting possibility of termination many of them had faced too often in the past. To return to PND just when their PGD cycle appears to have been successful is thus a difficult step for many couples, despite having been asked repeatedly, from the very beginning of their encounter with the PGD team, to imagine themselves in exactly this potential situation.

For Sally and Ben, who had never previously achieved a pregnancy (because of Ben's translocation, which causes oligospermy and consequent infertility), the question of whether or not they would undergo prenatal testing if Sally did become pregnant with PGD reproduced exactly the same dilemmas they had encountered when deciding to undergo PGD in the first place:

> BEN: The Professor, he asked us, didn't he? He said right, "I'm going to ask you this question now because you will never be in a better state of mind to answer this than you are now, because once you're pregnant it's a whole different ball game," which it is 'cause of the emotional difficulty of having the actual baby inside you and then having to decide. He said "If you, if we told you your child was Down's would you terminate?" That's what he asked us, didn't he?
>
> SALLY: Yeah. It's *such,* such a *difficult* answer to give! 'Cause you've got to think of everything. Not just yourself, you've got to think!
>
> BEN: [*interrupts*] But, he said to us, he said "I'm not—you know what you say now isn't going to be cast in stone" [*laughter*], he said, "but what you've got to think of is, you know, would you be able to terminate the child?" No. And we at that point in time we turned round and said "Well no, I don't think we would." Because

it's difficult, 'cause whereas friends of ours say "Oh yeah, no way—we'd terminate straight away." . . . But then you're thinking "Oh hang on a minute, you can get pregnant at the drop of a hat! You know, you can have children whenever you want to!" You know what I mean? "A bottle of wine and you can have a child!" Whereas *us*— [*laughter*]

SALLY: We can have bottles of wine and no children! [*laughter*]

BEN: If it's so difficult in the *first* place you think "Hang on a minute, it may be a Down's syndrome child but that may be the only chance you ever get to have a child!"

In answering the professor's question, then, Sally and Ben decided that they would not undergo prenatal testing, even if it meant a risk of Down's, in part because PGD might be the only way they could become pregnant at all. For them, the chance of terminating a much-desired child, even if the child did turn out to have Down's syndrome, was not really thinkable. Having spent so much time and effort trying to have a child, they could not imagine making a decision to terminate a child who "may be the only chance you ever get."

Other couples were less sanguine about the possibility of having an affected or disabled child. When asked what they would like to change about PGD, a number of couples mentioned that they would like the genetic testing to cover a range of diseases or even, as Rebecca called it, the (mythical) "whole genetic test." For these couples it was quite difficult to deal with the fact that, although the clinic would test for the known genetic condition, the PGD team could not combine this with tests for other common conditions such as Down's. If couples wanted to test for these conditions, they would have to undergo prenatal screening, and risk their costly pregnancy.

For some couples, the possibility that something else could go wrong with the embryo was almost impossible to imagine—so much preventive care had been taken already. When Melissa had an early miscarriage after her third PGD cycle, she and her husband Tony were shocked. Tony describes his struggle to continue to listen to the voice in his head warning him against the temptation to think that, after everything they have been through, "well what else can be wrong with it?":

In your brain you think you've gotta have the hard bit. When Melissa was pregnant in your head the first couple of three weeks is that if it hasn't got cystic fibrosis, it's gotta be working. But when we were back there, they said the reason you miscarried is that egg's probably—it's probably got something else wrong with it. And that was hard to take in then. You think, well what else can be wrong with it then?

159

Despite this experience, however, Tony and Melissa did not believe that more genetic testing at the embryo biopsy stage would be helpful. They had decided to undergo PGD because of a strong feeling that they could not knowingly bring a child into the world with cystic fibrosis, the genetic condition they carry. This conviction meant that any further information about other genetic conditions could also lead to termination. As Tony explains:

> Like there's no way! If they said to me, oh it hasn't got cystic fibrosis but it might have this other defect, or another one. If you knew that, it'd be really hard to say "Oh go and carry on anyway." I'd rather not know, would you? I'd rather they test for the cystic fibrosis [only].

This discussion indicates a core dilemma of PGD. How much genetic information is useful? What is the relationship of the genetic "known" to the genetic "unknown," or, more practically, how is the genetic information that PGD can provide contextualized by what it can't?

As these accounts demonstrate, one kind of genetic choice is not the same as another: there is considerable variation in the "answers" couples reach concerning "additional" genetic information. The choice to pursue PGD for a disease that is definitely present in one or both parents, in order to prevent implantation of an affected embryo, is very different from the decision to undergo prenatal screening to detect other genetic or congenital abnormalities that hopefully do not exist. Indeed, these choices are as emotionally opposite as they are clinically complementary—in yet another "paradox of PGD."

Conclusion

As this chapter shows, the "cautious optimism" that was expressed by the clinicians and scientists who began to practice PGD during the 1990s is also very much a feature of the experience of undergoing it, as described by couples whose hopefulness is "guarded" and whose determination is as essential as it is potentially undermining. PGD takes place in a "topsy-turvy" world of opposites in which decreased fertility is the cost of increased genetic control, and beautiful pictures of dead embryos recall both triumph and loss. Above all, what emerges from this portrait of going through PGD is the exceptional emotional resources necessary to get through it. The possibility of failure is prominently figured from the outset of treatment, and, while this is undoubtedly very difficult for couples to adjust to, and almost certainly accounts for the high dropout rate of PGD, it has the additional, and somewhat surprising, effect of also increasing their confidence in PGD as a route forward. The confidence required to go through PGD is thus in part established through the painful losses that bring potential patients to its door. Developing an "appropriate" level of engagement with the demands of PGD at the outset is not simply about acquiring a sufficient amount of information, technically or otherwise, or being fully informed about all the things that can potentially go wrong during a cycle. As is discussed further in the next chapter, PGD patients are required to ask difficult questions of themselves, and this creates an atmosphere of seriousness about the technique that is very noticeable in the transcripts of interviews with both patients and PGD staff.

In contrast to IVF, in which one of the most difficult challenges is to maximize hope for the ultimate outcome while simultaneously exercising hope control, in order to be prepared for the more than likely event of failure, the distinctive challenges of PGD involve decision making. These decisions, moreover, take place after the IVF phase of treatment has been completed, and they involve issues related to genetics rather than reproduction. This is one of the main qualitative differences between IVF and PGD—that IVF is oriented toward assistance to conception, while PGD offers assistance to heredity. The goal of IVF is a child, whereas the goal of PGD is, in a sense, the reverse, in that it is aimed at preventing some kinds of children being born.

IVF and infertility turn out to be a much larger part of PGD than many patients expected, and the difference between IVF for fertile couples and infertile ones is less dramatic than might have been imagined. IVF is not only a means of undergoing PGD, as in a kind of platform technology or prerequisite, but is a vital component of success or failure, since the chances of establishing a successful PGD pregnancy are primarily determined in the first instance by the number and quality of eggs that can be generated. From that point onward a steady attrition will occur at each stage of treatment, and chance will play as large a part as skill, experience, or effort in determining the outcome.

For all these reasons, and as explored in this and the previous chapter, PGD is not a technique clinicians readily recommend. While the PGD team is convinced of the efficacy of the technique, and know it can work, they are outspoken and even adamant about its drawbacks and limitations. This combination, of intense professional dedication to a technique that is clinically and scientifically arduous to perform but uniquely rewarding when it succeeds, creates the distinctively intense and serious atmosphere that pervades PGD treatment from all sides. It also creates the distinctive ambivalence toward PGD that is one of its defining features, as well as, arguably, being a paradigmatic feature of the "biosociety" more widely (Kerr and Franklin forthcoming).

In turning to the topic of "moving on from PGD" in the next chapter, one question is whether, at a certain level, going through PGD "works" because even when it fails it stands in both for children and for the exercise of parental duties, responsibilities, and obligations. In a sense, PGD enables a performance of conjugal unity through procreative activity *without procreating*. As we shall see, the pain and hardship of past grief and loss can be assuaged through PGD whether or not it succeeds, and the bonds established through the process of going through PGD can make it both more and less difficult to move on.

Chapter 5
Moving On from PGD

Steven: Well, that was Christmas. It was December last year that we lost the baby. And I, I think at that point I was very adamant that we weren't going to go through it again, actually 'cause it was just so painful! Physically, for Rebecca, and mentally for both of us that—I really, really didn't want to do it. But, I think once everything—in the cold light of day—once Rebecca's healthy, and I, I think the good news that we got out of that, is that we didn't know whether Rebecca could be pregnant. So—'cause you never had been so that was a sort of a ray of hope as well.

Rebecca: Yeah, I think it's however many—however many times they tell you "there's nothing wrong with you" or whatever, which I suppose they've always, they've always said to me, but still you know, I'm thinking, well . . . you know, I haven't . . . God, I mean you read in every magazine every day how it gets more difficult to get pregnant as you get sort of on! But you know, so we had, you know, we had a pregnancy and you know that was a positive thing. And, um, at the end of the day, if you, you spent our lives dwelling on the negatives, we'd have both topped ourselves by now! You got to look on the positive side of it really! And that was, that was one positive thing. [*pause*] I think it also, I think . . . how, how we dealt with the grief of that, certainly how I dealt with the grief of that, losing the baby, was that you try again!

Everyone who undergoes PGD will eventually have to end active engagement with treatment. For a minority, this will come as a logical progression following the birth of one or more unaffected children. For the majority of couples, however, "reaching the end of the road" turns out, like most aspects of PGD, to be more complicated than first imagined. Coming to terms with ending treatment means, for most couples, coming to terms with its failure, and facing the challenge of how to make sense of experiences that will remain with them for the rest of their lives. However, failure is also an incentive to continue, and, as Rebecca describes above, one way to deal with grief is to try again.

The quotation used as an epigraph to this chapter, from an interview with Steven and Rebecca, contains many of the key aspects of "moving on from PGD," including the pain of disappointment and grief following failed treatment, the way hope can transform ideas about the future, the complicated relationship between positive and negative aspects of treatment, and the unexpected ways that failure can revitalize determination to succeed.[1] All these factors contribute to the emotional difficulty of the deliberations through which couples make, and remake, the decision whether to end treatment or carry on. In this chapter, PGD patients' descriptions of their experiences, and the views of the professional staff who work most closely with PGD patients, are analyzed as a form of "reproductive accounting," referring to how couples weigh their odds or chances in order to reach a decision about continuing treatment, and how they account for, or explain, their actions. A different, contrasting, model of "reproductive accounting"—in terms of its accountability in retrospect—is the subject of chapter 6.

The model of "reproductive accounting" used to explore decision making about "moving on" from PGD helps provide some insight into why the issue of having and not having children in the context of assisted reproduction and heredity is so far from straightforward, and so much more difficult to manage than it at first appears to be. Since both PGD and IVF are oriented toward assistance in having offspring, it is all but axiomatic that they stem from the desire to have children. However, while it is also obvious that IVF and PGD are very different kinds of choices than fostering or adoption (or the choice to remain childless), these options are also less distinct than they might appear to be, as we shall see below. As interviews with PGD couples demonstrate, many of these "opposed" alternatives are significantly more interwoven, overlapping, and commingled than might be evident from the point of view of assuming that having "one's own" biological offspring is the whole point of undergoing IVF and PGD. By exploring the importance of alternatives to PGD, and in particular the role of ideas about *not* having children, this chapter returns to one of the central questions about reproductive assistance, which is the extent to which the very process of trying to have a child can in some ways substitute for having one.

Of course this is not to suggest that the one outcome is substitutable in any simple manner for the other. However, unlike IVF, PGD is not about

---

[1] All but two of the PGD patients interviewed for this study had experienced repeated PGD failure: six couples' cycles had ended early, either because of poor or negative response to medication, or because there were no suitable embryos to transfer. Two couples who had reached the stage of embryo transfer subsequently failed to become pregnant, while two other couples experienced brief pregnancies that miscarried or were terminated owing to other problems. Only one couple had succeeded in giving birth to an unaffected child through PGD, as is described at the end of this chapter.

promoting fertility but about preventing genetic disease. Hence, the desire to have children in the context of PGD is *inherently ambivalent*—in the sense that healthy children are wanted but children who will suffer a painful early death are not. As we have seen, PGD is contradictory, or paradoxical, in other respects as well. For most couples who undergo it, PGD is an impediment to fertility, and while you must be determined to succeed, you also have to be prepared for the four to one odds that you will fail. As this chapter demonstrates, these defining features of PGD are as prominent for couples exiting from their encounters with it as they are for those arriving at the PGD threshold to begin with.

The couples who gave interviews to this study held a wide range of attitudes toward the question of how to reach an end point to treatment, and had diverse sets of plans for "moving on." This was always a subject to which considerable thought had been given. Broadly speaking, there were two main framing devices for the question of how to reach an "end point" in treatment. In one frame, PGD was one of several equally desirable options. In the other, PGD was attempted "first" to see if it would work (and to maximize its chances of doing so), with a view to moving on to alternatives, such as gamete donation, surrogacy, or adoption, if it failed. For both younger and older couples, the clinically recommended three attempts were often imagined as giving PGD a "proper go." However, such plans could be disrupted by a number of events, as in the case of Rebecca and Steven—whose intention to end treatment following a traumatic miscarriage gave way to the decision to continue once an initial period of grief found its resolution in renewed hope. Not surprisingly, many couples found the question of when to end treatment confusing, in large part because of changes that affected them during the course of treatment.

This finding is widely reinforced in the literature on IVF treatment. Studies by Margarete Sandelowski (1993), Sarah Franklin (1997), Gay Becker (2000), and Karen Throsby (2004) have all explored the difficulties of defining an end point to treatment. As Thompson also notes (2005), the difficulty of anticipating what coming to terms with "permanent" failure is likely to involve is compounded by the disproportionate emphasis on successful outcomes in popular media representations of IVF, and by the frequent lack of any explanation for why treatment was repeatedly unsuccessful. An unfortunate "added cost" of unsuccessful treatment, as Throsby notes, is that the shame and secrecy associated with infertility may be intensified by repeated, unexplained, and emotionally traumatic IVF failure, leading, in some cases, to a desire to conceal, and thus keep further privatized and "hidden from view," what an ultimately "failed" outcome of treatment feels like (2004).

For couples whose PGD cycles repeatedly ended in failure, the main difficulty in determining an "end point" to treatment stemmed from changes to their initial expectations of what the "finish line" would be like—or where it would be. While many couples trusted the PGD team to inform them if they had reached a point beyond which further treatment attempts would be ill-advised, such a clear-cut end point is not clinically determined in the majority of cases, and even in the most apparently unlikely cases, clinicians will have seen the exceptional successes against all odds. The apocryphal tenth attempt at IVF that resulted in a healthy birth, or the patient who was dissuaded from further treatment only to succeed on the following cycle, are familiar enough from the world of IVF to make the idea of the "end of the road" elusive.[2]

Clearly, a risk of the "end point problem" is that couples may, through an inability to bring themselves to walk away from treatment empty-handed, having invested so much and having come to feel that abandoning hope is too painful, enter an increasingly desperate state that can ultimately cause them harm. (This "one more go" mentality is among many features of assisted reproduction—so often figured in terms of odds, chances, percentages, and luck—that render comparisons to gambling, horse racing, and card games ubiquitous.) From a clinical point of view, there is a strong sense of responsibility to prevent patients harming themselves, or others, by "going too far"—as the purpose of IVF and PGD is to alleviate suffering, not to increase it. However, clinicians often feel it is not their role to decide when a couple's "chances are up," or their hope has run its course, or their gamble has been lost—for these are complex personal and emotional issues, and at the end of the day, aside from the exceptional clinically "impossible" cases that prove the rule, it is the couple that has to decide how best, and how far, to proceed.

Shifting Goalposts

Elizabeth provided a vivid description of a typical "shifting goalposts" experience in an interview that was recorded shortly after her third failed PGD cycle. After two unsuccessful attempts at IVF, Elizabeth was referred to PGD. In her first two PGD cycles, no embryos had been suitable for

---

[2] The situation described in Rebecca and Steven's interview, of coming "so close" it inspires another go even after a decision has been made to stop, in part because stopping is so much harder after the added disappointment of a near miss (as opposed to coming nowhere near), again illustrates some of the instructive parallels between "reproductive roulette" and other forms of gambling. The logic, for example, of slot machines relies on exactly this "near miss as incentive" phenomenon. Slot machines are specifically designed to maximize the impression of a near miss, as the number of winning combinations, or "jackpots," is proportionally small, while the number of combinations that are only missing one component is very high.

transfer. Like many couples, she and her partner had set a limit of three treatments. However, on her third and theoretically final attempt, she and her partner Jack achieved their best-ever outcome, resulting in two "particularly good" embryos being transferred. Despite this auspicious indication, Elizabeth had recently discovered she was not pregnant. Not yet having been back to the clinic to discuss this failure at the time of the interview, she said that she and Jack did not yet understand why the cycle had failed, and were now considering a fourth attempt.

Elizabeth's description of her experience illuminates the problems of reaching the "last go" when it turns out instead to be an incentive to continue, and thus produces the problem of "not really knowing where we're going" or indeed which "road" she is on:

ELIZABETH: We sort of thought this was *it*, this was the *last* go. And we were sort of expecting probably the same thing to happen again [of having no embryos]. So then that would *definitely* be it. You know, there's no point having any more goes. But it hasn't! So we're sort of, we don't really know where we're going. [*little laugh*] We don't know whether— We're going to see them next week. So we don't know whether they're going to say "Oh there's no point having another go." Or whether they'll say "look you did really well this time, we don't know why it didn't work it's just one of those things. You should, you know if you want to have another go." . . . There's also the option of donor eggs, because mine aren't *fantastic*. And I do have somebody who is willing to give us some, but that again is another whole, ball park [*little laugh*]. Or just forgetting the whole thing.

SF: Yeah. So you'd like kind of drawn a line under it, that this would be your last go because (ELIZABETH: Yeah!) you needed to feel it wasn't going to go on indefinitely (ELIZABETH: Mm!), then because of the nature of this cycle, (Elizabeth: yeah, it's—) it's more of a question.

ELIZABETH: Yeah. I thought that I, that after this I'd sort of be at the end of one road and you know we'd have the adoption road, donor egg road, nothing. And we'd have to decide which one to do. And now there's this added sort of "well maybe we could have another go ourselves" you know, 'cause obviously that *is* the preferred option. [*little laugh*] You know if we do have a baby then ideally it would be nice if it was *our* baby, genetically. Um, so, yeah, we're a bit—we're sort of a bit in limbo now. We don't really know what our options are which—but you know, Jenny phoned on Monday, and I—she said, you know, "do you want to leave it? Do you want to come and see us?" I said, "Well I'd really like to see you quite soon. Just because I, I like to know where I'm going." I said, "Just so that I know what all the options are, and then we can take a few months out and, you know, mull it over and see what we decide to do." So that's where we are. [*little laugh*]

Elizabeth's prediction that she would be "at the end of the road" after three PGD attempts made sense at the outset of treatment, but not *after* she made unprecedented progress on the third go, changing the stakes and renewing a sense of hope. A sense of having overcome what was formerly the most significant obstacle has left Elizabeth and her partner in a different position in relation to their future: their near miss has thrown them back into limbo. Rather than convincing her that, after three failed goes, PGD is not the right "road" for her and her partner, Elizabeth's most recent experience has led her to feel she might be closer than ever to success. What is notable is that she has been *doubly deprived*: she has lost the closure that a more definitive failure would have provided while also being deprived of success. This deeply ambivalent point in treatment, lost in the uncertain and possibly treacherous territory between just enough and too much hope, can feel impossible to navigate, or even to assess.

As Elizabeth states, PGD is "just one road" for people in her situation. Being at the "end of one road" does not mean that there are not other roads to go down: as she lists, there is "the adoption road, donor egg road, [do] nothing [road]." But as this metaphor also emphasizes, she is at a crossroads, faced with a very difficult choice and with limited means of evaluating her options. What is making her decision difficult is, not only the level of uncertainty about the many factors affecting her treatment in the future, but also the fact that she has already gone back on one of the earlier decisions that was supposed to be her guide (i.e., to stop). This means that for Elizabeth both the criteria for making her decision (its content) and the process of making it (its form) have been altered, and she feels at a loss to know how to proceed. Her decision is to wait for more information from the clinic, which will help her decide how much effort in pursuit of PGD is "worth it" and also how much she and her partner can take.

Counting Cycles

Elizabeth and Jack's decision to try PGD three times is a common one, and is the number most routinely recommended by clinicians (see also Throsby 2004). Many couples set a limit of three cycles in advance deliberately in order to avoid overinvesting in treatment and being drawn into the dreaded "treadmill effect" of simply carrying on because they are unable to stop. Joanna and Christopher were embarking on their second attempt when they were interviewed and had, like many couples, made a definite decision to return to "trying naturally" after three cycles if they were unsuccessful. They had decided this was an appropriate number not only because they were concerned not to exhaust their life savings on PGD but also in an attempt to retain a sense of control over their lives and to facilitate their own physical and emotional well-being. The specter of "desperate" women subjecting themselves to over twenty cycles of IVF haunted Joanna, and was something she was determined to avoid. In her account of PGD, she expresses her strong desire to have another child but also to protect herself from becoming "desperate" for a baby:

> JOANNA: I *don't* want to be somebody who um. . . . I've heard so many stories of people who have just spent their whole life savings having—not particularly PGD but IVF—and have had like 22, gone through it 22 times! And because I just know this past year, it has—it *does* take over your life because it's very difficult to plan anything because you know, all the time you're looking at your diary thinking well if my period's then, and then count 3 weeks and then that'll have to happen, so during this month, I'm going to be going back and forth to London but because you don't know *exactly* when my periods going to come and *exactly* when the dates are going to be. It's *really* difficult to actually plan anything, and things like jobs. I'm doing this job at the moment but it's not *really* what I want to be doing but— So it's like everything else is sort of put on the back burner while you do this, and the thought of just spending the next 5 years or whatever doing that just—has—
>
> CHRISTOPHER: Yeah.
>
> JOANNA: I don't want to be *desperate*, I don't. . . . If it doesn't work, I don't—I will try very hard—I don't want to be—'cause I've met women as well who are just very, very desperate to have a baby. And I can understand that, 'cause deep down I really, you know, I *am* because I *really, really* do want to have another child but um, well obviously . . . ! [*laughter*] The ah—it's, I don't want to be a desperate woman, 'cause—because— [*pause*]

At this point, Christopher provided his own analysis of their decision to undertake no more than three cycles. For both him and Joanna, three cycles constituting a year's treatment would leave them both "basically emotionally and physically knackered."

> CHRISTOPHER: Well we said from the outset that we just wanted to at least know we've *tried*. We didn't want to keep going until it worked, we just wanted to have a go and we've always said three. I

mean three goes has always been what we've discerned as being available, I think what we decided. . . . We hadn't made a conscious decision that it would be three, it just seemed—that seems to be the right number. 'Cause that's basically—you've been trying for—if we *do* that, it's a year and a bit and that's, that's like as Jo said, everything goes on hold. You don't plan—you don't take holidays, trips away, you—it just takes over.

JOANNA: I also don't think it's particularly good for your body to do that on a regular basis. After however many years I'm sure that—

CHRISTOPHER: For a year to, a year to 15 months, after that, if you're basically emotionally and physically knackered, that bit seems to be the time to say to well let's stop you know, enough's enough.

Three cycles are going to be enough for Joanna and Christopher to feel they have given PGD a proper chance to succeed and thus, as Christopher says, "at least to know we've tried." This number is definite and premeditated ("we've always said three") but not for any reason, other than that "it just seemed to be the right number"—so that it is at once arbitrary and "right." Above all, it is meant to be protective. In a sense it matters less what the number is than that there is a number, so that the treatment does not "take over" and there is an agreed-upon limit at which point "you know, enough's enough."

This logic is rational and prudent. It is cautious, responsible, and protective, in addition to being clinically recommended. Like many decisions taken in the context of PGD, it expresses a sense of responsibility, awareness of risk, and obligation on the part of others as well as oneself. "Three" is a sensible, obvious "logic" and provides a cautious limit, but many couples who plan to undergo only three cycles find that implementing this plan becomes more complicated as treatment progresses, and a different logic eventually takes over. This is not because patients have become "irrational" but because there is more than one rationality operating in the context of assisted conception—indeed there are many. Even the apparently simple arithmetic of counting up to three cycles can prove perplexingly elusive in the context of PGD. Which cycles count as "real" cycles? Should couples include cycles that were terminated early? Or cycles in which they decided to go ahead with embryo transfer despite embryo quality being so poor that their chances were virtually nil? These are some of the questions, often rather technical, that many couples never consider before beginning treatment and setting a limit of three attempts, as the following conversation with Kirsten and Scott about their upcoming cycle demonstrates:

SF: And so will this be your *second* cycle that's coming up now?

KIRSTEN: Well, it's technically the *third*, because the first one— (SF: Yes) that, hopefully will go from the full—

SCOTT: [*interrupts*] That never went through, that never went through the process, did it?

KIRSTEN: —from start to finish, this will be the, the third, yeah.

SCOTT: The first one was drugs only, to find in examination that no, this isn't going to work, they knew from that examination. Then the *second* one, drugs and process went through complete procedure but failed, and this will be the *second* drugs and process through.

Here, then, Kirsten and Scott have decided to call their third attempt at PGD their second cycle, because their first attempt had to be halted owing to Kirsten's response to the ovarian stimulating drugs. From this experience emerged a distinction between a "full" cycle, which "goes from start to finish," and a "drugs only" cycle, which "never went through the process" and therefore doesn't count.

Drawing a Line

The complexity of the kind of "reproductive accounting" described above, in which couples weigh the odds and make choices, only to discover that the rules of the game have been changed, is driven partly by the fact that the outcomes of PGD cycles often improve over time, as clinics acquire more information about the woman partner's response to ovarian-stimulating drugs and can "fine-tune" her cycle with greater control.[3] Even repeated failed cycles can become a source of encouragement to continue beyond the previously stipulated three attempts because they are diagnostically cumulative, and in their own way mark a certain kind of progress (Franklin 1997). Having come closer than ever to a successful cycle in her third attempt, for example, Elizabeth no longer felt that three goes were sufficient to "exhaust" the PGD option:

ELIZABETH: So, you know, a few days ago it was like "no, I don't think I can do it again." But already I know that I'm changing my mind and I know that if they say we can have another go, a few months down the line I'm gonna want to.

[3] It is important to emphasize that although the language PGD couples use emphasizes their identity as couples, each partner has very different roles, the primary physical burden and risk being born by the woman partner. Like other language in the context of PGD, such as "patients" and "treatment," "couples" can be criticized as euphemistic, as in "couples undergoing treatment," because PGD is neither treatment nor undergone by a "couple" (it is the woman partner who is the subject of the vast majority, although not all, of the medical interventions). However, a notable feature of PGD, like IVF, is that the unequal, gendered, reproductive labor of treatment is subsumed, and often deliberately downplayed, under a rhetoric of shared reproductive aspiration—much as Thompson describes in her account of the "ontological choreography" of IVF (Thompson 2005, esp. chap. 6).

SF: Yeah, 'cause it's *very* hard to stop when you've had sort of the best (ELIZABETH: Yeah!) go you've ever had! (ELIZABETH: Yeah!) You've gotten further.

ELIZABETH: Yeah. So you know, I'm not ready to give up. And I don't think that I can—I don't think that we can sort of consider adoption or anything else until we've exhausted *this* line. I think you have to draw a line and come to terms with that before you start looking at something else.

The passage of time played a prominent role in Elizabeth's decision. Initially, even after her third near-miss cycle, she was certain she would not continue treatment. But as the days passed, she looked back on her third cycle differently, sensing that she was beginning to change her mind. Instead of the intensity of recent disappointment, her view shifted toward the future: she now feels she cannot draw a line under what she has been through, or be through with it, until she has "exhausted" the PGD option. In order to give up, in order to put PGD behind her and move on to adoption or other alternatives, she has to come to terms with PGD—a task she feels is likely to involve at least one more attempt. This experience replicates the difficulties of counting to three: often it seems you are ending when you have only just reached a reasonable starting point.

In sum, it is hard to stop when you feel you've only just begun. After three failed PGD cycles, Melissa and Tony describe how gaining *more* knowledge makes them feel *less* confident about how to proceed:

TONY: If we could start again now.
MELISSA: With what we *know.*
TONY: And what *they* know, it'd be much better.

Here again, in the "topsy-turvy" time of PGD, things shift unpredictably. Whereas for Elizabeth, the time of the future can emerge only when she has successfully put PGD behind her, Tony and Melissa want to turn the clock both forward and back to make their most recent attempt their starting point—which is what it feels like it has been, but only because they were so far behind to begin with.

Reproductive Accounting

Eighteen months prior to their interview, Anne and Daniel had made a decision to end PGD treatment, which they explained in terms of *chances and percentages.* They had reached this decision following three failed cycles, including one in which a (very rare) contamination occurred in the laboratory, meaning that the embryos could not be implanted. After three cycles, Anne and Daniel were told by the clinical staff that they could not expect more than a 20% chance of PGD ever working for them. After so much disappointment, and having reached the end of their funding, these odds did not seem "worth" another go. Through a careful weighing of their chances, Daniel articulates a form of "reproductive accounting" in which odds and percentages are reckoned alongside all the costs and expenses involved in PGD:

> I mean, I think if our chances would have been the same as everyone else, I think, it's obviously still a big thing, £4000, whatever it is to go through another cycle, it's quite a lot of money, you know, you can't just go on indefinitely, paying £4000 every six months or whatever, that sort of, that affects your lifestyle if you're going to do that. But then to be told that your chances have been halved, or, whatever, put down to maybe 20%, you think well, it's still costing us the same amount of money, you know. If it had *halved* the cost, then we might have thought, oh well, you know, but it's— So we sort of, we sort of took everything in our hands, you know, the cost, the fact that our chances had been reduced, 'cause we don't produce good quality embryos, and I think that sort of package said no, really.

Complicating the language of chances, percentages, and costs in Daniel's accounting is a sense of relative value: "if our chances would have been the same as everyone else," Daniel begins, it is "obviously still a big thing" to pay four thousand pounds per cycle. Since "you can't just go on indefinitely" paying so much money, the clinic's estimate that their chances are, in a sense, twice as expensive per cycle, provides an occasion to sum up the whole "package." The idiom of "cost" clearly applies to more than money, and the decision to reach an end point is described as putting Anne and Daniel back in control.

The variables for such reproductive accounting are difficult to gauge, and considering them all in balance is also difficult because it involves weighing such disparate factors. For some couples, as discussed in chapter 3, financial issues are an important criterion: not using all one's life savings on reproductive treatment is a significant factor in decision making. Age can also provide a cutoff date. Belinda, for example, said she had "earmarked forty." This "drawing of a line" was not only age related but also, as Belinda explained, derived from her perceptions of the impact of treatment on her relationship with her partner, Nicholas. Similarly, Nicholas was concerned about Belinda's health:

> SF: In terms of imagining how the future will go, you said that you'd be prepared to do it until you're forty. Is it important to you to have an end in sight, to be able to cope with all the different aspects of it now? Like an either/or way of thinking about it?

173

BELINDA: Yeah I think, um, it *does* help to, to be able to draw a line under it when you stop, cause I, otherwise I would just probably keep going and, and then that would make Nicholas unhappy, because it, it's *very* disruptive in our lives, it *really* is.

NICHOLAS: It's not so much in our lives, it's the effect it's having on Belinda, on Belinda's *body*. Um, you know, the fact of the matter is if you're making the body produce twenty times the number of eggs, what's it doing to her ovaries? And, you know, it, it, the whole treatment process is getting more sophisticated, but it's pretty crude really.

For Belinda and Nicholas, the "costs" of PGD are accounted for in terms of protecting their future relationship and Belinda's future health. These efforts to balance the costs of PGD are also being weighed against the state of treatment, which, while it is "getting more sophisticated," is assessed by Nicholas as "pretty crude really" in a reference to his doubts about its safety.

Against the prospect of hoped-for success, the determination to persevere, the sense of obligation to do what they can, and the other factors in its favor are the cumulative "costs" of PGD: it may not be very healthy, it is expensive and time-consuming, it "puts your life on hold," it is disruptive to your relationship, it is physically and emotionally draining, it is a "gamble," and the odds are stacked against you from the start. Given these disadvantages, it is inevitable that many couples ask themselves not only why anyone would find it difficult to put an end to PGD treatment but indeed why anyone attempts it to begin with. However, as shown in the previous two chapters, and in discussion of a successful case below, there are numerous, compelling, and very obvious reasons why people attempt PGD. But PGD is not simply an arduous option that is *also* a compelling one: what some of the more complicated hesitations about ending PGD treatment suggest is the somewhat worrying possibility that PGD is attractive *because it is arduous*. As we have seen, couples who are willing to attempt PGD, and who decide to carry on with treatment despite being made aware of its considerable drawbacks, must be understood to be seeking more than success. In other words, a willingness to "exhaust" all the options, no matter how trying, is a risk worth taking in part because it may provide a sense of having completed a course of action, *whether or not* it succeeds.[4]

---

[4] This possibility, that conjugal aspirations are served by undertaking a course of action *because* it is arduous and thus, as in the classic tradition of romantic love, provides an obstacle to be overcome that affirms (or destroys) mutual commitment, would be the place to explore in more depth how biological reproduction is implicated in the performance of conjugality "required" by heteronormativity. Although such a task is beyond the parameters of

Again, this is a "risk" of PGD treatment of which clinicians and genetic counselors are well aware. Why should trying to have children necessarily be easy? After all, everyone knows that having them is difficult, if also rewarding. Is it possibly important to consider whether choosing PGD is choosing to undertake a trajectory toward having children that expresses shared parental desire and enacts this desire as shared effort, hope, and commitment, in a way that has value in itself? If the possibility that this affirmation of parental intent may become an end itself is part of PGD "treatment," how should this be evaluated? Is there a sense in which trying to have a child stands in for having one, so that stopping trying is, in that sense, a double blow because it takes away both a hoped-for future and the hopefulness of action in the present? What would the conditions be that would enable the best decisions under these circumstances?

A possibility is that a missing component in discussions about these issues is the role of *social expectations*.[5] If many couples are expressing both a desire for a child and a desire to conform to the expectations of "society" and their peers that they have children, how might we revisit Mary Warnock's question about "what kind of society can we praise or admire" in which we "can live with our conscience clear"? Whose conscience needs to be clear, and of what? What is the relationship between the often highly privatized feelings and emotions of couples undergoing PGD, the empathetic but strictly confidential relationship they have with the PGD staff, and the wider society of which they are part? These issues are especially prominent in terms of how couples consider alternatives to PGD, and in the question of *the social accountability of PGD*, which is the focus of chapter 6.

Other Options

The question of how normative expectations—for example about gender identity, heterosexuality, conjugality, kinship, and family—structure attitudes toward "achieved" conception is crucial, as the role of technology

---

this book, it is a general theme of early feminist work on reproduction that has, as yet, received little critical attention (see further in Throsby's 2004 account of IVF as a means of "negotiating normality").

[5] Many of the studies of infertility and IVF have emphasized this point, including those of G. Becker (2000), Crowe (1985, 1990), Franklin (1997), Sandelowski (1993), Thompson (2005), and Throsby (2004). In all these studies, the processes of naturalization and normalization involved in IVF have been emphasized. Interestingly, and as was noted in the introduction, PGD does not "imitate" nature in the way IVF does, and "normalization" is also differently operative in relation to the genetic component of PGD than in the strictly reproductive purpose of IVF. These issues are discussed further below.

might be seen to offer greater capacity for these to change, but instead what we often see is the hope that technological assistance will repair, replace, or restore a normative state. New reproductive and genetic technologies are often described with dramatic labels such as "designer baby," however, precisely because the transformative power of technology is in tension with the tendency for ideas of the "natural" or "biological" norm of reproduction to remain the same, that is, to proceed along a narrowly defined path involving two parents whose offspring equally share their reproductive substance, thus embodying their union and affirming the bond between them. This tension, which might be described as the conflict between "bionormativity" and technological assistance to reproduction, has structured much commentary on the social dimensions of new reproductive and genetic technologies, and has been most eloquently explored by Marilyn Strathern, who noted in 1992 that it was technological assistance to reproduction that made the hybridity of reproduction—its dual identity as both social and biological—newly explicit and that, in doing so, created new ambiguities about the meaning of "biological facts": "There is a new ambiguity about what should count as natural" (1992b, 19), she noted, claiming also that technological assistance to human reproduction would have effects on cultural knowledge. "The more facilitation is given to the biological reproduction of human persons, the harder it is to think of a domain of natural facts independent of social intervention. . . . [This] will not be without consequence for the way people think about one another" (1992b, 30).

The question of how new technological possibilities enter into, disrupt, confirm, or simply coexist alongside more traditional expectations about the "naturalness" of reproduction takes some interesting turns in the context of PGD, which, unlike IVF, is not an "imitation of life" but, interestingly, an artificial form of inducing biological death. It is thus a form of assisted conception through which a couple can affirm their shared conjugal bond by preventing their reproductive substance being equally embodied by a child "of their own." Instead, as an expression of reproductive commitment, obligation, and responsibility, PGD responds to a new form of reproductive desire, which is to exercise a form of genetic contraception that will prevent harmful genes being reproduced. This, consequently, is, as Strathern describes it, a "hybrid" form of reproduction, the implications of which, as this book argues, are only beginning to be fully understood.

One way to look at the relationship between "technonormativity" (faith in scientific progress to improve the human condition and willingness to pursue technological options on behalf of an improved future) and "bionormativity" (what is considered to be biologically normal) is thus to explore how PGD couples consider other options besides PGD—

options that do not involve the biological norm in which procreativity physically confirms and "completes" conjugality, or options in which this norm has to be refashioned.

PGD is "officially" described as a specialist alternative, as one among several other options, and, by implication, as suitable for some people rather than others (HFEA and ACGT 1999, 3; http://www.doh.gov.uk/genetics/pgdprinciples.htm). As noted earlier, the people for whom PGD is seen to be most suitable are couples who are at risk of giving birth to a severely genetically affected child and who are unwilling to undergo prenatal screening, or couples who cannot produce a viable pregnancy owing to chromosomal translocation (for whom prenatal screening is not an option).

However, even these couples always have other options besides PGD. The pressure to undergo "the PGD route," then, is always relative: it is strongest ("our only choice," "not a choice") when the other options (such as adoption, fostering, gamete donation, "trying naturally," prenatal screening, or childlessness) are *least acceptable*. If the desire to have a biologically related child is strongly preferred over adoption or gamete donation, then PGD is the "only option." However, for many of the couples interviewed, PGD was seen, both by choice and by necessity, as one of several *equally preferable* options. Whether PGD was seen as the only option or as one of several did not affect the determination with which it was pursued. A couple who would have been happy to adopt if PGD failed was nonetheless likely to feel the same desire and determination to undergo PGD as a couple for whom adoption was an unacceptable option. Similarly, there was no evidence that couples with a preference for PGD found ending treatment harder, or easier, than couples with a more explicit sense of PGD as one of several possibilities. These different reasons for undertaking PGD all had different implications for how couples reached an end point to treatment, and how they moved on from PGD, as we shall see in the following sections.

Adoption

All of the PGD patients interviewed for this study were asked about their views of alternatives to PGD, and in every case it was clear that these had been carefully considered and discussed in depth. Moreover, and as mentioned above, it was evident that among the cohort of PGD patients interviewed for this study, PGD was not strongly preferred over other options, such as adoption, and neither was it chosen in order to ensure that a child would be biologically related to both parents.[6] To the contrary, although all the PGD couples interviewed were clearly aware of the normative expectation that they would have a child "of their own"—meaning genetically related to them both—this was not described a predominant factor in their choice of PGD.

For example, Kirsten and Scott had investigated adoption in the past and, although they had some strong reservations about open adoption in particular, were planning to return to considering adoption seriously if their upcoming PGD cycle was unsuccessful. In the following extract Scott explains the positioning of this option as the next decision he and Kirsten will need to make:

> SCOTT: [*sighs*] I think we've got to the stage now, we've actually accepted that if *this* doesn't happen, then that's it, we're going to shut the door on it. And we've got to decide, um, really, our only other alternative is gonna be adoption. . . . I think as Kirsten said earlier, we're taking each step at a time. . . . but we always said that we would continue up until we were forty and then after forty, we'd *then* consider other alternatives but I think that's a decision we've still got to make. . . . I think we've got to put the lid on it at *some* point. When you consider the treatment that you've gone through, yeah? It's got to come a time when we've just got to accept, well this ain't gonna work. So yeah, we'll have to see. . . . We know there's options and we know that there's gonna be a decision at the end of it.

In this example of "reproductive accounting" several factors are being calculated including age, the need for limits, and the order of the options that will be pursued. Like many other couples in their situation, Kirsten and Scott described a series of options roughly mapped out ahead of them, which provided the basis for a step-by-step decision-making process. They were not opposed to adoption, although, like many other couples, they were aware of some of its perceived drawbacks. However, as Kirsten said explicitly, a preference for her "own" child was not about

---

[6] Since this study was not representative, it is not possible to generalize on the basis of the small number of patients interviewed what percentage of PGD couples have investigated alternatives, or to what extent. Moreover, such data are largely missing from the expanding literature both on new reproductive and genetic technologies and on "alternative" families, or families by choice (Judith Stacey 1996, 1988; Franklin and McKinnon 2001a, 2001b; Carsten 2000, 2004). There is thus an obvious opportunity to bring together work on adoption, lesbian and gay families, single parenting, and so-called blended families with work on family formation via the NRGTs. A major study intended to do just this, titled "The Genetic Tie: a Cultural Account" (Franklin, McNeil, and Roberts 2002), which sought to compare ideas about the role of biogenetics in family formation, was refused funding because of the perception that the study design was itself a sign of overinvestment in the concept of "geneticization" (which it was in fact designed to investigate on the presumption suggested by the research described here that it would be *less* prevalent than imagined).

wanting a child to be genetically related to her: "if I'm not genetically related to my child I don't feel that's anything, I think it's how you bring the child up."

Other couples had already initiated adoption procedures.[7] Rebecca and Steven, for example, had begun to investigate this option with their local social services authority.

> REBECCA: So . . . now we've come home and contacted all the . . . local authorities and had a chat. We had a *really* nice chat actually, and are just waiting for the other ones to come round. So, sort of, if this next one, sort of our third [cycle] . . . doesn't work. . . . It's not something I can happily say we're not going to try again. And I don't think you ever really know how you're going to feel, *after* you've been through it and it's been successful or been unsuccessful. We'd *certainly* love to—ah, we've *always*, always said that we'd think of adoption anyway, didn't we?
>
> STEVEN: Yeah! Even before we—even before were married you know, I just think it was one of those things that we could do, that, you know, would help children. And I *love* children and you know I hate seeing children being treated very badly. I think it's something that anybody that's got, you know, a nice house and a good job can give back to society! It's very sad to think [in] this country that we've got a big adoption problem. . . . It's heart breaking!
>
> REBECCA: Very sad.
>
> STEVEN: Maybe that's my calling!
>
> REBECCA: Maybe it is! And maybe that's why I— [*laughter*] Maybe we're fighting against the inevitable! [*laughter*]
>
> STEVEN: Yeah! We'll have several adopted children! [*laughter*]

For Rebecca and Steven, making these initial inquiries strengthened their knowledge that, as Rebecca put it, "PGD . . . is not our only chance of having a family." They also expressed their desire to adopt both in personal terms ("even before we were married") and in terms of the wider society ("something that anybody . . . can give back"). Steven's comments notably combine the most commonly expressed component of a desire for children, of love for them, with a sense of social duty and obligation, and with an explicit awareness of social problems. The sense of a calling also informs this account, demonstrating how much a desire to adopt and a desire to have "children of your own" can have in common, despite often being seen to be opposed.

---

[7] At the time of this study there were approximately sixty thousand children in care (effectively, "wards of the state") in the whole of the United Kingdom. Approximately five thousand children are adopted annually in the United Kingdom.

Gamete Donation

Gamete donation was another option that some couples had considered or tried. In the following extract from an interview with Rebecca and Steven, they discuss their views about gamete donation in terms of their own, and their partner's, reproductive hopes, as well as the all-important question of "nature versus nurture." For both of them, the difference between nature and nurture feels arbitrary, as do the distinctions between adoptions and gamete donation. This fascinating account of "where children come from" demonstrates their ability to switch back and forth between options that, far from being opposed, look increasingly interchangeable:

> STEVEN: Well, PGD isn't our only option. You know, there's still IVF through donor—
> REBECCA: [*interrupts*] Yeah—I don't want to do that.
> STEVEN: —insemination, which is what [he would prefer]. I can't see what difference that [sperm donation] makes, actually!
> SF: [*To Rebecca*] Would you rather adopt?
> REBECCA: Yes, I would, yes. I think doing it *that* way [through sperm donation], it's almost a selfish, it's because *I* want to have *my* only child, bear my *own* child.
> STEVEN: But *I* want *you* to have had that chance as well!
> REBECCA: Right, well there's two things. I think it's a selfish thing I want to bear my own child when there's lots of children out there, who I could look after. And it's this selfish thing that they've got to have some sort of genetic connection to *me*. The major thing is that at the end of the day, I wanted to have *your* child. You know, that's, that's the big thing, I think, *if you're a couple, that's what you want*! And at the end of the day, it wouldn't be, well it would presumably. . . .
> STEVEN: Well, it would! . . . See I just don't see that! [*laughter*]
> REBECCA: They're all—they're all sort of shades of the same thing really, aren't they? And I'm a great believer in nurture as opposed to nature. I do think that is, although I'm sure there's a bit of both, but I really do think that the sort of um— [*pause*] that a child, if, if it wasn't genetically related to Steven, . . . it would still, have um, his, you know, their personality would be based on his personality, . . . I do actually believe that's true. And at the end of the day what you want, and why it is his child, you want it to be like him. Maybe it won't look like him, but you want it to be like him as a person. And I think that if we get that, I accept that, but I think it would be the

same with an adopted baby! I think if you were lucky to get a baby you could adopt, and get it from an early age, you would do exactly the same thing in a way! Nurture, not nature. And if, you'd also be [able to] . . . give a child a home that it wouldn't have had! So [much the better].

STEVEN: But what I *still* would like—

REBECCA: Mm!

STEVEN: —would like my wife to be pregnant and have a child . . . with somebody else's sperm, should I say! [*laughter*]

REBECCA: You're not suggesting I just go out and . . .

STEVEN: No, I'm not! [*laughter*]

REBECCA: Good! [*laughter*]

Significantly, Rebecca and Steven's conversation about IVF, adoption, and gamete donation emphasizes "likeness" more than biology. Consistent with both Helena Ragone's analysis of surrogacy couples (1994) and Cori Hayden's account of lesbian parenting (1995), the origin of reproductive desire is described in terms of the conjugal bond and the desire to fit in with normative expectations, rather than with the idea of biogenetic continuity. In Ragone's famous phrase, Rebecca and Steven's imagined children are "conceived in the heart," rendering the precise means of their arrival less a matter of biology than of determination and a unity of purpose.[8]

What is evident in Rebecca and Steven's comments is, once again, the embeddedness of the PGD option within other options, and the extent to which these are explicitly substitutable for one another ("I can't see what difference it makes actually"). Rebecca's comments also demonstrate an important slippage between what an individual or couple wanted and *what they perceived they were expected to want*, as in Rebecca's comment that *"if you're a couple that's what you want."* In fact, as her comments demonstrate, it is not entirely clear "what you want" when you want a child. These views about biology, family, kinship, relatedness, and reproduction also reference wider ideas about social justice and responsibility, as when Rebecca describes as "selfish" her desire "to bear my own child when there's lots of children out there I could look after." Her emphasis on the importance of genetic connection is contradicted by her view that "they're all sort of shades of

---

[8] The question of gamete donation raised by Rebecca and Steven was raised by a number of couples, some of whom had received offers of egg donation from siblings. Although none of them had gone ahead with this option, these offers provided an important "backup" option when PGD began to seem unlikely to be successful.

the same thing" and that "I'm a great believer in nurture as opposed to nature."[9]

Genetic Relatedness

For Rebecca and Steven and indeed many couples who were interviewed for this study, thinking about using donor eggs or sperm raised explicit questions about the importance of genetic relatedness. Like the participants in Charis Thompson's study of gestational and egg donation arrangements in a California infertility clinic, couples who imagined future reproductive possibilities using donor gametes were able to "strategically naturalize" their relationships (Cussins 1996, 1998; Thompson 2001, 2005). This term, part of the "ontological choreography" through which parents remake relationships, identity, kinship, and personhood, refers to one of the signature elements of the "world of assisted conception"—namely the ability to reconfigure unusual reproductive arrangements *as inherently similar to "natural" reproduction* in the effort to normalize them and make them more acceptable (see Franklin, Lury, and Stacey 2000). The plasticity of the *ART (Assisted Reproductive Technology)–imitates–life* imperative (which might, in contrast, have been expected to be quite rigid) is evident in Thompson's study of couples who were using family members as surrogates to carry their children, where she describes women who argued that what was most important about their own bond to the potential child was the *genetic* connection (embryos made from their own gametes were transferred to the other woman's uterus) rather than *gestation*, and contrasts this position against that of other women, who were using donor eggs from close friends but carrying the fetus themselves and who argued the reverse— that it was gestation, not genetics, that created the most meaningful bond. Thompson thus documented among the participants in her study a strikingly flexible ability, both as individuals and as couples, to "strategically naturalize" technological assistance to conception—thus making it possible to normalize and legitimate what could otherwise be seen as deviant or stigmatized means of reproduction.[10]

[9] These equivocal comments, which exemplify the "hybridity" of ideas about kinship and procreation in contemporary British society described by Strathern (1992a, 1992b), are discussed in more detail in J. Edwards 2000. See also Bouquet 1993; Carsten 2000; Franklin and McKinnon 2001a.

[10] The extent to which both kinship and biology have become much more plastic, or flexible, and the relationship between these two phenomena has been the subject of an increasing literature in anthropology as well as other disciplines. See Carsten 2000; Edwards et al. 1999; Franklin and McKinnon 2001a, 2001b; Thompson 2005.

Although none of the couples in this study had actively pursued surrogacy, a number raised this as a potential solution for them if PGD was unsuccessful. Elizabeth, for example, had received an offer from her sister not only to donate eggs but also to carry a child for her and Jack. "Strategically naturalizing," Elizabeth describes in the following extract how the fact that her sister has "come from the same place I have" makes sister egg donation seem like a more attractive option than adoption. Unlike adoption, where "you're never quite sure what you're getting," using her sister's eggs would be "the closest thing" to having her own baby.

SF: You said it's quite important to you to have a child that's *yours*, genetically.

ELIZABETH: Yeah, [*uncertain tone*] well, I er, yeah I, I think it's more important to Jack, for some reason. I'm quite willing to adopt, if this doesn't work. I want to—I want to give this my best shot and, and not feel that I gave up too soon. But if it doesn't work, I want to—either I'll go with my sister's eggs because I sort of feel that if I'm gonna go use eggs, donor eggs, she's come from the same place I have, so it's still the same genes, if you know what I mean. It's not *me* but it's the closest thing I've got. You know, and quite feasibly the baby could look like me. 'Cause actually my sister's little boy looks like me not her [*laughs*], which is quite strange. Um, so there's *that* option. And although the baby wouldn't genetically be *mine*, I would have carried it, and it would almost be. Um, but if that isn't, you know, an option, 'cause we're not sure about that, um, *I* would be quite willing to adopt. Jack's not so sure. He feels, he, he's worried about bonding, and he said you know, like you're never quite sure what you're getting. [*little laugh*]

Elizabeth and Jack had not reached the stage of taking this route at the time of the interview, as Elizabeth stated that she needed to have "come to terms with the fact that I'm not going to have my own child before I can think of having someone else's." She had, however, put serious thought into considering what it would mean to carry a child made from her sister's egg and Jack's sperm and had decided that, in contrast to asking her sister to be a surrogate mother, this would be tolerable because she would "bond" with the fetus in utero.

ELIZABETH: It would genetically be *his* and my sister's, which is a bit weird [*laughter*] but, you know, but then if I carried the child for nine months, it would be my child. She just said you know "I'm just giving you something that I'd throw away every month." [*smiling*] So she said, you know and I, I wouldn't, I couldn't do it if she had to carry a child 'cause I think that would just be so *unfair* on her. I'm—I know that if I said, to her, "I want you to carry" then she *would*. But I couldn't ask her to do that. I couldn't ask her to carry a child for nine months then give it to me. I think that would just be too much, too much to ask of somebody.

SF: Right, yeah. But if she went through egg recovery and she donated her eggs, and then if they fertilized and they would transfer it to you then it would be more like giving birth to your own child?

ELIZABETH: Yeah, yeah. Well, you've done the whole *nurturing* thing before it's born, so. You know, she wouldn't have bonded with her eggs, [*little laugh from both*] if you know what I mean!

The use of donor gametes in reproductive medicine was receiving some attention within the British media at the time of this study. Under the recently established Human Rights Act, court cases were initiated to overturn the legal right to donor anonymity, arguing that children's human rights are violated by the withholding of information about genetic fathers.[11] The couples who spoke about using donor gametes were aware of these debates, and many spoke of concern for potential children regarding whether they would know their genetic parents.

Gamete donation, surrogacy, and adoption all raise different issues about the significance of genetic connection—a topic about which interviewees had conflicting views. This range of views on the importance of "nature versus nurture" was reflected in the diversity of alternatives that had been considered and in the depth at which they had been discussed. As in other studies by anthropologists of the meaning of blood ties and genetic connection, in particular the work of Jeanette Edwards (1999), Janet Carsten (2004), and Judith Modell (1994), what was also striking was the degree of *substitutability among and between what are commonly assumed to be quite different kinds of choices*. This same effect, *of nominally opposed meanings standing for the same thing*, was also evident within accounts of particular alternatives, for example when Elizabeth described being a great believer both in nurture and in the genetic tie. Indeed, Rebecca's comment that "they are all more or less shades of the same thing" characterized much of the content of discussion about alternatives to PGD.

---

[11] Legislation was passed prohibiting egg and sperm donor anonymity in Britain in January 2003.

Trying "Naturally"

For many couples the most obvious, although never simple, option when PGD failed was to return to trying to become pregnant "in the usual manner" and then use prenatal testing. As this option was felt to be intolerable at the time of taking up PGD treatment, its reappearance when PGD failed is significant.

Anne and Daniel's reproductive accounting after three failed cycles led them to decide to try to conceive a child "naturally" and to use prenatal testing for SMA. In describing this decision, they describe PGD as "just another option" on a long list of possibilities. Despite their experience of failure, they remain "very positive" about PGD and the St. Thomas' team. Indeed, they remain as positive about PGD as they were two years ago when, speaking at a national conference for SMA families, they advised other couples to consider PGD:

> DANIEL: We're still very positive, I mean, when we did our speech on PGD two years ago, our final line was "It's not worked for us, but we still remain very confident in the team at St Thomas' and, you know, we'd advise it to anyone who wants to try it." Because for us it's just another way that you can choose—there's lo[ts?], there's a *few* choices: you can, you know, remain childless; you can just try naturally . . . and whatever you get, you live with; you can go by the CVS route; or, or now, it's a new one, the PGD route. So it's just another—
> ANNE: It's another option isn't it?
> DANIEL: It's another option, yeah.

Even in their own case, they are not ruling out more PGD in the future:

> SF: Yeah, and, and as you said, it's not an option that you've ruled out necessarily, it's just not—
> ANNE: No, it's just not, it hasn't felt right for us for the last year and a half, um, and we've decided to, um, try naturally but without any success. [*little laugh*] Um, but I, I think we've, I've got some sort of fertility problems anyway but, um, but we've decided that we're sort of going to take that route and we go for the test in 10 weeks and, um, you know, see from there and *maybe*, you know, if we are lucky enough to conceive again and then we have the test and then the unthinkable happens, and we have a termination, then that might push us back in the direction of PGD again, I suppose.
> DANIEL: I think the other thing as well is that we're also thinking that maybe in a couple of years time, if we went back to see Professor Braude, he'd say "look, the success rate has gone up, we've never

had contamination since, we can control embryos better so that quality's better, everything's looking a lot rosier than two years ago," we might say "Oh right, yeah, you know, things have got a bit better, we might have another go then, yeah."

For Anne and Daniel, "trying naturally" is both a way to move on and a way to manage the disappointment of PGD failure. Significantly, they now imagine their reproductive future hand in hand with scientific progress, and will consider PGD again if it improves. The ability to move in and out of PGD over time is seen to increase their reproductive options, which may therefore, and somewhat ironically, improve with age, if, as they imagine, things look a bit "rosier" in a few years' time.

Living without Children

Since the perceived aim of PGD is to help couples have children, its alternatives are frequently defined in terms of other ways to have children, but there is of course the option of not having children at all—by PGD, adoption, or any other means. While it may appear obvious why this option is least often encountered in discussions about alternatives to PGD, this is also why the option of childlessness has become one of the most overlooked topics in debates about new reproductive and genetic technologies (c.f. Gillespie 2003; Letherby 1994, 1999; McAllister and Clarke 1998).[12] In every generation there are, of course, many people who never have children and are perfectly happy not doing so, but this is now a more acceptable option, in particular for women, than may ever historically have been the case. In Britain, for example, some of the highest statistics ever recorded of numbers of women who have never had children *pertain to the same baby-boomer generation that was the first to benefit from IVF.* Thus, interestingly, despite the fact that increased, and indeed historically unprecedented, rates of voluntary or "chosen" childlessness and the opportunity of having children via IVF are coincidental phenomena affecting the same generation, they are rarely discussed together, or as related.

Instead, the most common association between childlessness and IVF is the perceived link between "delayed" maternity and *involuntary* childlessness, for which IVF then becomes the "solution." Clearly, however, one of the major factors affecting both women's and men's decisions to have children, biological or otherwise, is the much-publicized tension

[12] According to National Office of Statistics data in 2002, Britain has one of the highest rates of childlessness in Europe, with twice as many women over forty remaining without children compared to the previous generation.

between work and family life, for which few either innovative or well-resourced solutions have been devised. From this point of view, both the desire for children and the choice to remain childless may have in common a degree of ambivalence about the conditions available for raising children. It is almost certainly this ambivalence that has led many women to "delay" childbirth into their thirties or, indeed, "permanently."[13]

Despite the fact, then, that ambivalence toward having children is increasingly prominent and is a significant factor affecting demand for IVF, the link between IVF and *not* wanting children is rarely explored. Indeed IVF is routinely linked not only to the desire for children but to a "desperate" desire for them. In general, both IVF and PGD are depicted as synonymous with *unequivocal* desire for parenthood. But none of the research on IVF supports this view, revealing instead, as this study does, a much wider range of attitudes toward having children than either wanting them no matter what or wanting them biologically or not at all. Nor is this surprising given that most couples who undergo IVF or PGD will fail, and will need to have some alternatives in mind. What this study, like others, has shown is that even among couples who express a very powerful determination to pursue PGD there is a strong awareness of other options, and a self-conscious effort not to be too single-minded in their pursuit of any one of them.

In addition to other possibile ways of having children, PGD couples who were interviewed also spoke about the possibility of remaining childless. Katherine's comments, in which she expresses both a "definite" interest in adoption and the possibility of continuing to have a "great life" without children, are typical of the ways in which a familiar pattern reasserts itself, in the form of one set of possibilities that can be substituted for another:

> KATHERINE: I mean lots and lots of our friends have said, you know, "What about adoption would you consider it?" You know we, we, we've been through it with loads of our friends because I think it's something that they would definitely do if they were in our position, but no I think that we um—we've got a great life, and you know children will be *fantastic* but it doesn't happen for everybody.
>
> SF: Yeah. And you feel that um, you know, there's a load of different possibilities in the future. . . .

---

[13] The relationship between educational level and childbearing among women is particularly striking, as it is prominently among the highest-educated women that rates of childlessness have risen most dramatically in the postwar years. This phenomenon has received scant attention and yet is directly relevant to IVF due to the disproportionate number of IVF users who are highly educated.

KATHERINE: . . . That's right. And you know, at the moment I'm still young enough to keep going a little while longer! [*laugh*] Going on for 6 years now, so— But I think we've got another 6 years, realistically, so I think you know we'll use that time and, you know, and if it doesn't work out in that time, it's not as if we're going to say— look back and think "Oh god we should have—"

For Katherine, then, the issue is not so much having children at any cost but feeling that she and her partner will be able, *whether or not they have children*, to continue to enjoy their lives together. Although children would be "fantastic," Katherine expresses a sense of acceptance of the different future that may be ahead of her in her statement that "it doesn't happen for everyone."[14]

The Void of Childlessness

One of the most negative associations with childlessness is of emptiness and even pointlessness of existence. As the Warnock Report claimed in sympathetic but somewhat uncritical language: "there is for some couples the view that the desire for children is the result of a deep, unbidden, biological drive which cannot be suppressed." Indeed, there are many versions of the "essential" importance of having children— ranging from genetic determinism to religious fundamentalism. However, it is also the case that the rise of new reproductive and genetic technologies has significantly transformed ideas about how children can be had, and this radical twentieth-century transformation of the "facts of life" has been accompanied by a significant shift in the kinds of family forms that are legally and socially recognized—from lesbian and gay families to "reconstructed" families to transnational-adoption families (Carsten 2000; Franklin and McKinnon 2001b).

However, despite the fact that IVF and PGD have played important roles in changing the ways in which families can be formed, they are often associated with enabling couples to have the "same" kinds of fami-

[14] Significantly, however, her ability to accept that children may or may not be part of her and her partner's future is linked to the importance of having tried PGD. As she says, "it's not as if we're going to . . . look back and think 'Oh god we should have—' " Here, again, the question of the extent to which the mere existence of some reproductive options exercises a subtle pressure on some couples to undergo them, at the very least so they can say they tried everything, needs to be considered. This reference to fear of future regret is very common in accounts of both IVF and PGD, and although it can be a positive way to accept failure (at least we tried everything and will not look back with regret), it is also a potentially worrying incentive (we have to try because we might later regret not having done so).

lies as "everyone else." In other words, IVF and PGD, while associated with whole new kinds of offspring, such as miracle babies, designer babies, and savior siblings, are at the same time repeatedly normalized, naturalized, and contextualized within the narrowest and most traditional definitions of family, gender, and kinship—as the biological nuclear family. Indeed it is striking that although IVF inaugurated an entirely new form of achieving a pregnancy, which is entirely unlike what has since come to be known as "spontaneous" or "old-fashioned" pregnancies, it is primarily associated with nature, the "facts of life," conventional gender roles, and traditional family values.

One of the criticisms of IVF is that it makes the experience Katherine describes above, of simply accepting that having children is not something that happens for everyone, more difficult. In a society that places a high value on individual responsibility and effort, as well as on scientific and technological progress, there is inevitably a pressure on couples to pursue every avenue of possibility. As Gay Becker observes in her study of US IVF consumers, there is pressure to be seen taking action to remedy infertility and not remaining "passive" in the face of adversity (2000). The possibility that conception can be achieved through IVF, like the perceived option to improve on hereditary outcomes made available by PGD, produces *a new form of social responsibility* as well as new choices. Put together with the pressure on couples to "complete" their marriage through reproduction, to provide grandchildren and heirs, and to express themselves through maternity and paternity is the additional imperative to strive to make progress in the face of adversity (G. Becker 2000; Franklin 1997; Throsby 2004; Strathern 1992a, 1992b).

Indeed, for some (although not all) of the PGD couples interviewed, one of the most disturbing aspects of repeated PGD failure was the sense of moving closer and closer to an unknown space of a potentially childless future. Despite (or because) of being only in their late twenties, Sally and Ben were already very concerned about what this "void" might feel like. For them, having children had always been what they expected to do with their lives.

BEN: It is the single most, biggest thing that most human want in their lives innit, so?

SALLY: It's what you aim for isn't it? From when you're little.

BEN: [*interrupts*] So people are going to jump at it [reproductive technologies]. If that gives me more chance then I'll go that way.

SALLY: It's a void as well I think when you get the—when you come to the point that you're going to stop IVF, I think you going to. . . . I think there'll be a void there. Because at the moment it gives us—

we're actively doing [something] to try and have a family. And with the PGD we're trying it even *more*! You know, we're going that one step further and I think if we say now okay enough's enough, we're going to have a void that we're not trying, and especially when they start coming out with all the new sciences, which they will do, it's inevitable! I mean the last thing we heard about is these stem cells with Christopher Reeves isn't it? That to me is *absolutely* amazing. To make somebody walk again! If they can do that then they can do a hell of a lot more in the way of, well trying to bring children into the world.

The sense that is conveyed at the outset of this extract, of children being the "biggest thing that most humans want in their lives," and of this expectation having always been a part of life, "from when you're little," is of powerful societal expectations shaping a sense of identity and belonging. In his reference to people who will jump at the chance to use new reproductive technologies, Ben invokes the "obvious" logic that links difficulty in conceiving with technological assistance. In her frequent references to the "void," Sally refers both to the potential absence of children *and to the absence of trying for children*. Indeed the void she specifically refers to is *of stopping IVF*. What is equally emphasized in her comments, then, are the importance of children *and the importance of trying to have them*. In referring to "a void that we're not trying," and to PGD as "trying . . . even more," the emphasis is, significantly, as much on the importance of trying as on the importance of children.

As Sally and Ben go on to say, if they do not succeed with PGD, they will move on to other options:

SF: So, how much do you think about the future, I mean, since you are so young [*laughter*] and obviously you could have a lot of goes at this and, you know, obviously hopefully it will be successful, but do you think about the future in terms of the possibility that it might just continue not to work and then and then what would you do?

SALLY: Well, I mean [*little laugh*], we've always said we'd have three attempts. This is our third attempt. We've always said that if it doesn't work, emotionally and physically we can't keep going through it, 'cause it is a strain. This is our third one in just over a year and we're going to look at other options. We said we're going to look at perhaps donor insemination or, um, adoption.

BEN: Adoption.

SALLY: But we've also said that perhaps later on in life after a couple of years we'll probably go *back* to the PGD 'cause it's going to be, probably be slightly more advanced and it's—

BEN: [*interrupts*] We'll have a break.

SALLY: —something I wouldn't like to say is that we're going to shut the door to it, that'll be it. I think just have a break now and pursue other, other directions. I think there's donor insemination, there's adoption, there's all sorts, isn't there, really?

As for many couples in this study, Sally and Ben's knowledge that technology will improve makes it difficult to "shut the door" to PGD.[15] Hence, in their account of their future, Sally and Ben will continue to weigh up options as they get older and the technology potentially improves, and they may eventually "go back to PGD" when it is "slightly more advanced." This likelihood of technological improvement is thus one of the major reasons it is difficult for Sally and Ben to stick to their limit of three goes, to give up on the idea of trying harder, or to abandon their determination to succeed. This is evident in Sally's subsequent explanation that she would like to continue with PGD indefinitely, despite the fact that she and Ben recognize the costs of putting "life on hold" for so long, have definitely set a limit of three attempts, and are already considering other options:

SALLY: But then having said *that*, I would have as many attempts as I could if we weren't paying. That probably sounds very shallow, but if the, if the government were going to fund it for us then yeah, brilliant we haven't got the worry of where to find the money from for every attempt! I'd like to have a couple of attempts a year until we got *somewhere,* because I'm, you know, quite confident that it would actually *get* somewhere in the end! I mean it *is proven* isn't it, with the IVF, that after so many times, so many cycles, 8 or 9 cycles, women are more likely to get pregnant! But it's a gamble isn't it? And that's why I think you've got to reach a point for our relationship's sake really. And for our social side of life, because we don't have a social side of life, if. . . .

BEN: [*interrupts*] No! [*sighs*] We haven't, have we, had since we first started it, have we?

SALLY: All we talk about is, babies and making babies! [*laugh*] And it gets quite boring to other people I imagine! [*laughter*] But it's still a thing to put life on hold for things like going on holidays 'cause you can't afford it 'cause you're always try and think about the money towards IVF. And we said 3 attempts and then pursue other options. Knowing that later on in life we *could* go back to PGD with it being more advanced perhaps.

Such contradictory impulses are intrinsic to the experience of PGD, for the same reasons they are so prominent a feature of IVF treatment

[15] For similar data in the context of IVF see Franklin 1997, 185.

(Franklin 1997, 194). Like PGD, IVF is undertaken to provide resolution to a reproductive dilemma (by producing either a child or a sense that the couple have "done everything" to try to have a child), but for some people it undermines the possibility of such resolution (because the process of "coming so close" can intensify the desire to continue, and because it turns out to be difficult to feel you have "done everything"). It is also because IVF is constantly evolving and improving that couples' reproductive lives become intertwined with the hope for scientific and technological progress. As a result, many IVF couples do not reach a point in which they feel they have "done everything": instead they reach a point where they are too old or too exhausted (physically and/or emotionally) to continue.[16]

Making New Connections

As noted at the outset of this chapter, everyone who undertakes PGD will at some point share the experience of having to reach an end point to treatment. Most commonly, patients will have to do this without having achieved the ultimate success of PGD—a child who is born free of a specific genetic disease, or in spite of an interfering translocation. However, as the data in this chapter also show, a child is not the only potentially positive outcome of PGD. It is also possible to derive satisfaction and a sense of shared achievement from having given PGD your best try, from those aspects of the technique that have succeeded (such as producing good embryos), or from being "free" to move on to something else.

The experience of PGD is consequently not entirely negative or unproductive even when it fails. Although the experience of failure can be painful and disappointing, and indeed devastating and overwhelming, its meaning changes over time, and is more complex than the simple opposition between success and failure implies.

In addition to confirming a couple's sense of having "tried everything," PGD can also produce new forms of connection both within families and to other patients. Because of the familial nature of some of the conditions that bring couples to PGD, the experience of going through a number of failed cycles can produce new kinds of dialogues and sharing within families. For example, the experience of discovering a translocation and tracing it back through family members provided for some interviewees a new explanation for their family's reproductive history, changing their relationships with parents and siblings. For one interviewee's

---

[16] "That *it is not possible* to 'try everything' is the realisation with which many women terminate their relationship to IVF" (Franklin 1997, 193, emphasis in original).

mother, the investigation of her daughter's and husband's translocation provided an explanation for her own miscarriages that had occurred thirty to forty years ago, for which she had always blamed herself. Although encountering this new explanation late in life was "very, very hard for her," it altered a long-term perception of hers that she was responsibile for those reproductive problems.

The experience of having been part of a team, working on the cutting edge of reproductive biomedicine, was also important for some couples, especially those for whom previous experiences had left them feeling badly treated, misguided, or abandoned. More than one of the couples interviewed had been the subject of television or media coverage, and some had become active in patients' groups as a result of their experience of PGD.

PGD Joy

However, the most important outcome of PGD is the potential for success of an almost indescribable kind. In the case of Natalie and Andrew, the only couple interviewed for this study who had succeeded in having a child through PGD, an introduction to their lively and engaging two-year-old daughter provided the occasion to see a very different PGD from that described so far in this chapter and those preceding it.

Like other couples who were interviewed, Natalie and Andrew spoke very candidly about the pain of losing first one and then a second child to SMA—both of whom died before reaching two years of age. In particular the suffering their children underwent while dying, and their inability as parents to do much except provide constant palliative care, were described in agonizing detail. Natalie and Andrew also described each child's personality and charm in loving memory of their brief lives, and with a sense of dignity that was still punctuated by the anger of profound grief and loss.

It was while caring for their terminally ill children that Natalie and Andrew became involved in the Jennifer Trust, the main patient organization for SMA. It was at a meeting of the trust where they first heard about PGD, which they described as being "the ideal scenario." Andrew reported their early thoughts:

> This is just what we want, this PGD. It's perfect for us, its perfect for SMA, you know. It's what we've been waiting for really.

Natalie agreed, adding that she and Andrew "felt so lucky that we were young enough and it's come about." Their experience of PGD treatment was overwhelmingly positive, and Natalie described it in simple, direct, and optimistic terms:

They wait 'til they've developed so many cells, take one cell out, test it and then hopefully you have some to put back in. And then basically they pop them back and off you go and keep your fingers crossed.

Having achieved a successful pregnancy on their first cycle that led to the birth of their healthy daughter, Emily, Natalie and Andrew were planning to return to the clinic for another cycle at the time of their interview.

The extremity of the contrast between Natalie and Andrew's descriptions of their first two children's deaths and their obvious joy in their two-year-old daughter's vitality was made particularly vivid when Emily entered the room toward the end of the interview. Unlike most other PGD couples, who distance themselves from the idea of a "designer baby," Natalie and Andrew were unperturbed by these associations, joking about their "Gucci baby" and her already-evident talents:

NATALIE: She thinks she's four!

ANDREW: She's a complete handful! [*laughter*]

NATALIE: . . . She's just like life, in't she? And the thing is, I just laugh 'cause like, she's so advanced! . . . When she came out, people who don't even know about her past, they say "God, she's unbelievable! She's been here before that one!" And I go 'mmm!'" . . . She came out like bright as a button!

For Natalie and Andrew, as for the staff who work on the PGD team and for the wider community of reproductive biomedicine of which they are a part, the proof of the importance of PGD is most powerfully evident in children such as Emily, who embody the benefits of modern technological intervention into reproduction and heredity. From the point of view of an experience such as that of Natalie and Andrew, PGD "makes sense" in the simplest way possible—as a viable alternative to an otherwise traumatic process of beginning a pregnancy unsure of its outcome, often while still grieving over the death of one or more previous children.

The challenging aspects of PGD are not only difficult to see but can appear almost incidental beside the affirmation of its success that a child such as Emily, and her parents' happiness, so vividly proclaim.

Emily and her parents, however, also demonstrate the extremity separating couples whose success is almost indescribable from those for whom making the best of failure is their only option. This is a substantial challenge to PGD staff, as the preceding chapters have shown, as well as to policy makers, and to "society" as a whole, which is the subject of chapter 6 and the conclusion. One of the major issues arising from reproductive

biomedicine, after all, is the relationship between the reproductive desires or successes of individual couples who choose to undergo the new
techniques and the wider questions that a technique such as PGD poses
for everyone else. The following chapter turns to this wider question of
how to "account" for PGD not only in terms of how it affects individual
couples, or the professionals who work with them, but also in terms of
how affects the wider society of which they are part.

Chapter 6
Accounting for PGD

Tony: I don't look at them as doctors. Do you?
Melissa: No.
Tony: . . . Not 'cause they're not professional or nothing, just because they're really, really good to you. They never ever speak down to you. . . . And they involve you with everything, which is great for me!
Melissa: They do explain everything and if it hasn't been successful then when we go back there the file's out, they go through everything—what they think went wrong and what's this. They're all really good.

The preceding chapters have introduced a range of divergent perspectives on PGD—from patients, clinicians, scientists, policy makers, journalists, bioethicists, philosophers, and other commentators as well as from the media, government agencies, and the voluntary sector. Together, these accounts of PGD comprise an ethnographic archive, through which a collection of disparate and conflicting understandings becomes a sociology-of-PGD database. The contrasts, alignments, and comparisons between and among these versions of "what PGD is all about" evoke Meg Stacey's (see introduction) mandate to describe and apprehend the "manifold and complex" nature of social life, its complexity, and in particular the "interrelatedness of acts and effects" (M. Stacey 1992).

In compiling a description of this kind it is habitual and expected that the result is an account of diversity, plurality, and complexity—for this is the "problem" of society that is readily to hand for legislators, regulators, policy makers, and government advisers (Strathern 1992a). However, to do more than simply depict the endless variation of perspectives through which to imagine "science and society" or "the social impact of the new genetics," it is necessary to produce a more precise formulation of the analytical goal. Thus, in attempting to provide *an account of PGD*, this chapter also explores the ways in which PGD is *called to account*, or *made accountable*. Hence, in addition to an account *of* PGD, this chapter asks how PGD is made accountable "back" to the wider society. By this means, the question of "PGD and society" can, in a sense, be asked both ways 'round.

The Social Life of PGD

The elusive question of how one makes visible the sociality or "social dimensions" of a technique such as PGD can be approached, as a whole trajectory of social studies of science has already done, by putting the social into science and technology by demonstrating their internal social dynamics—such as the traditions, rituals, hierarchies, and conventions with which all scientists are familiar (for an excellent example from medicine see Atkinson 1995, and for an overview see Franklin 1995a). Another approach is to analyze scientific knowledge production in terms of the wider socioeconomic and national-political influences that have produced genomic science in the form we find it today (for several examples of this approach see Goodman, Heath, and Lindee 2003 and see Marks 1995, 2002, as well as Jasanoff 2005). Turning to questions of social welfare and social justice, it could be asked: Who are the new genetic technologies for? Who benefits from them? Who loses out? What are the forms of power and inequality that are channeled through geneticization, genetic determinism, or what Dorothy Nelkin and Susan Lindee call the "new genetic essentialism" (1995, and see Duster 2003).

The model of accountability explored in this chapter combines elements of both of these approaches—examining not only the social forces within PGD that influence the form of its practice and application but also the external forces, such as patient demand, in relation to which PGD protocols are adjusted, revised, and altered. This approach, then, emphasizes *exchange and interaction*—which are also built into the "two-way" model of accounting being used here. Finally, in its emphasis on exchange this model of accounting returns to the *social contract* that grounds the Warnock Committee's recommendations and to the key principle guiding the "Warnock strategy," which was that *in exchange* for permitting a limited amount of embryo research the state would assure its strict regulation subject to the very highest standards of public accountability. In asking "What kind of society can we praise or admire? In what sort of society can we live with our conscience clear?" Mary Warnock posed a question about *feeling, judgment, and belonging* to which she offered a solution of *tolerance, compromise, and regulation*. As argued in chapter 1, PGD played a key role in the public and parliamentary deliberations shaping the outcome of legislation in 1990, and could even be described as their "tipping point," when, at the penultimate moment, its

197

clinical success was revealed on the eve of the parliamentary vote. The enactment of the Human Fertilisation and Embryology Bill in turn inaugurated the Authority, which has worked hard to keep PGD "in step" with public opinion. How well has this regulatory process dealt with the often difficult questions posed by PGD, such as "savior siblings," aneuploidy screening, and the question of what counts as a "serious" genetic disease? What does it means to "keep in step" with society when even the medical profession itself is deeply divided about such issues?

To keep in step, be it marching or dancing, requires close observation, attention, and coordination, as well as practice. Learning to keep step with the rest of the troop is one of the primary components of basic military training because it instills individual discipline while creating a sense of a whole that is greater than the sum of its parts. In a sense, this describes the kind of "social contract" that marks the "British way forward" with techniques such as PGD and IVF, as well as with human cloning, stem cell technology, and biobanking. There needs to be a firm hand disciplining the troops, and if they can maintain discipline and order, they will make much faster progress—to everyone's benefit.[1]

PGD and Accountability

As previous chapters have shown, the range of positions on PGD spans the gamut of "deeply held opinion" on "matters of life and death"— from outright opposition to passionate advocacy. There are additional conflicting perspectives *within* these positions, about which many of the participants in this study were both reflective and articulate, and ambivalent as well as passionate. By emphasizing the overlaps, gaps, and tensions between and within different and conflicting views of PGD, the aim throughout *Born and Made* has been to present these different accounts in as much depth and detail as possible—and to foreground the ways in which the most difficult, challenging, confusing, or worrying aspects of PGD are negotiated. In turning to the question of what these negotiations can reveal about the social character of new reproductive and genetic

---

[1] Britain's National Health Service continues to operate through forms of hierarchy and discipline directly shaped by its military origins and subsequent conversion into a public welfare service at the end of World War II in response to the enormous casualties of that war. For example, nursing training in the United Kingdom remains distinctively military in its codes of discipline and practice. The blunt and unsentimental attitude familiar to patients within the NHS is often unsettling to visitors from countries where health services are oriented toward "customer satisfaction." The NHS is the largest organization in Europe and is recognized by the WHO as one of the best health services in the world.

technologies, this chapter attempts to analyze this negotiation process by naming its specific interactions.

In using the idiom of accountability, the intention is neither to seek to assign responsibility in a legalistic manner (liability) nor to invoke the auditor's impartiality so favored in the controversial pursuit of transparency-by-quantification associated with the so-called "audit culture" (Power 1999; Strathern 2000). For reasons that are explored further below, accountability is not used here in the sense of a moral discourse of greater transparency, objectivity, quality assurance, or any of the other harbingers of what Michael Power has described as "accountability creep" (2004). Rather, the question of accountability is posed in its *interrogative and explanatory* senses, by asking how an enumeration or reckoning (an "account") can be made of various aspects of PGD, and also how a process of questioning can be pursued (to give an account of, to account for).

As noted in the introduction, many mainstream representations of PGD depict it as essentially unaccountable. Either because it is highly technical and scientific or because it is tied to individual desires and choices, PGD has been seen to epitomize the failure to achieve adequate public accountability in the domain of new reproductive and genetic technologies. This chapter takes the opposite view, that in many ways PGD is highly accountable to its users and to the wider society—often in somewhat unexpected ways. As noted earlier, many of the prominent assumptions shaping the terms of public debate about PGD introduced in the opening chapters turn out to be starkly opposed to the views and opinions of PGD patients and members of the PGD clinical and scientific community. In contrast to the idea of choosy parents "designing" their offspring to have blond hair and blue eyes, PGD patients describe a strong sense of obligation to steer a responsible course away from avoidable harm. Although highly determined, PGD patients are not consumer-driven, competitive, or ambitious would-be parents seeking perfect "made-to-order" offspring. The image of PGD offering "too much choice" and "too much control" conflicts with the repeated depiction of PGD by patients and practitioners as error-prone, "crude," and three times more likely to fail than to succeed. Far from offering unprecedented genetic possibility, PGD is frequently described as the "only choice" by couples who see themselves as having no alternative. In contrast to the frequent emphasis on genetic selection and enhancement, PGD couples are often left with a very limited choice of embryos, if any, to transfer at the end of any given cycle. The optimistic anticipation by some PGD specialists of multiplex diagnostic assays capable of detecting "all known genetic disorders" may suggest an improved future for PGD, but the technological obstacles to such a quantum leap in diagnostic

199

accuracy do not appear to be lessening over time. Thus, in contrast to its image as a technology that is "racing ahead," the pace of PGD appears comparatively slow. Indeed, PGD intractably remains a daunting technical procedure in no small part because it relies on IVF—a technology about which questions are increasingly raised in terms of its effects on embryo quality. Against such technical odds, the image of PGD as a runaway technology, which is outpacing society's abilities to cope with its ethical challenges, makes little sense to the experience-near inhabitants of the "topsy-turvy world" of PGD.

More importantly, the "designer baby" view of PGD as a technology outpacing society's ability to restrict or control its use overlooks the long history of critical assessment and public debate that has surrounded it for more than two decades, particularly in the United Kingdom, where, as we have seen, PGD was instrumental in the creation of new laws governing human fertilization and embryology. The questions, worries, concerns, and hopes regarding PGD among those who work most closely with it have been a crucial influence shaping the way the technique is carefully restricted in its use, and this demonstrates a serious and concerted attempt to make it as accountable as possible to public scrutiny. As noted in the introduction, it is impossible to work in the field of PGD without experiencing an *intensification* of concern about the difficult choices raised by new reproductive and genetic technologies. Far from being distant, remote, or antisocial, the people involved most closely in the "world of PGD"—be they clinicians, patients, policy makers, or journalists—inhabit an intensely emotional world that is driven by complex combinations of obligation, determination, responsibility, and hope. These powerful affective forces must be reckoned with in any attempt to account for PGD, as in any effort to make PGD socially accountable.

PGD and the Risk Society

One of the most unexpected findings of this study was the considerable effort made by PGD clinical staff, at both of the fieldwork sites in London and in Leeds, to dissuade their potential clientele from using PGD, despite their commitment to its efficacy and their faith in its potential to alleviate suffering. While this cautious and preventive[2] attitude toward PGD was somewhat surprising, even more remarkable was the positive

[2] The "preventive" attitude toward PGD can be defined as the effort to protect patients against having disproportionate expectations of its capacities, and to discourage patients from using PGD until they are fully aware of its drawbacks and shortcomings.

value prospective patients attached to such "negative" depictions. Rather than describing their diminished confidence in either the PGD technique or the PGD team, patients who gave interviews to this study expressed the opposite, describing their *reassurance* and *increased confidence in success* after being told how overwhelmingly likely it was that they would fail. In their view, what might seem to be an overemphasis on "what could go wrong" made the PGD team, and the technique itself, *more accountable.*[3]

Such a finding both complements and contradicts current policy advice in the rather broadly defined area of "science and society." One of the primary conclusions of the House of Lords Select Committee on Science and Technology Report on the "crisis" affecting public attitudes toward scientific expertise in the UK was that "suppressing uncertainty is bound to diminish public trust and respect" (House of Lords 2000). This view derives from the prominent sociological argument that the late twentieth century ushered in an era of unprecedented skepticism toward scientific expertise and authority, leading to a desire for scientific uncertainty to be acknowledged rather than suppressed.[4] In what the German sociologist Ulrich Beck (1992) has denominated "risk society," scientific progress always brings with it new uncertainties and new risks—the very means that are used to create more control inevitably draw attention to the extent to which this control is eluded (c.f. Caplan 2000; Lupton 1993). Couples who feel more confident about PGD despite having been told it is highly error-prone, uncertain, and likely to fail do so because such admissions of inadequacy by professional experts *convey accountability*, and thus strengthen the sense of being treated with respect. The explicit references PGD patients made to their appreciation of having had their faith in the abilities of medical science knocked down a peg or two, and their testimony to the reassuring effect this had on them, can be understood within the model of the risk society, which presumes a skeptical and doubting public who value honesty, directness, and the "blunt" truth over unreliable spin and irresponsible hype. This "knowing" attitude, referred to by Anthony Giddens (1991) as the hallmark of "reflexive" modernity, is particularly prominent among educated, comparatively privileged, cosmopolitan elites.

Such sociological arguments are contested by those who see them as expressions of postmodern pessimism, such as the philosopher and

[3] It is always the case with small-scale studies such as this one that a different group of interviewees might well have provided different responses. For example, as noted earlier, PGD "refusers" and "dropouts" would inevitably offer different perspectives. However, as noted in chapter 2, qualitative data do not need to be fully representative in order to be *indicative* of important dimensions of the PGD experience that may be quite widespread.

[4] Social-scientific research on the risk society, and in particular the work of British sociologist Brian Wynne, played a significant role in the House of Lords Report.

parliamentarian Onora O'Neill, who argued in her high-profile BBC Reith lectures[5] of 2002 (nationally broadcast and subsequently published in the midst of this study, www.bbc.co.uk/radio4/reith2002) that uncertainty and trust are not by definition incompatible, and that increased information and transparency can be the enemy of trust. According to O'Neill, people want information that will enable them to make decisions based on sources they can check, from people they can trust, and many modern medical encounters epitomize the problems of how to do this (2002b, 77).

The model of accountability suggested by the data in this study offers an interesting case of the problems of accountability and trust O'Neill foregrounds. The repeated observation that patients find acknowledgment of uncertainty *reassuring*, and the corresponding hypothesis that PGD patients *value uncertainty* because it increases their trust in the clinical team, is not, importantly, an endorsement of uncertainty *itself*, but of the kind of accountability suggested by O'Neill's model, in which trust is built up through openness to critical interrogation. From this vantage point, increased trust can be generated both through acknowledgment of uncertainties that are by definition partly unknown *and* through explanation of them in terms of what is known.

These contrasting aspects of uncertainty are described by O'Neill as a "layering" of information (see below). As a member of the House of Lords Stem Cell Committee, and as a parliamentarian actively involved with patient support groups, O'Neill has become an increasingly prominent figure in debates about the "new genetics" in Britain. As an academic as well as a parliamentarian, O'Neill has made a number of increasingly influential arguments about the limits of informed consent in this field of medicine (2002a, 2002b), challenging the notion of autonomy derived from bioethics.

It was in her capacity as a scholar and a public intellectual that O'Neill agreed to be interviewed for this study in 2002 at Cambridge, where she is principal of Newnham College. In a discussion of what kind of information is most useful to patients undergoing procedures such as PGD, O'Neill introduced the concept of "extendable" information:

There's no answer to the question "how much information do people want"? And there are good logical reasons [for this] that have to do

---

[5] The Reith lectures were inaugurated in 1948 in honor of the first director general of the BBC and are designed to commemorate his view that broadcasting should enrich national and cultural life by inviting a leading expert in a field to deliver a series of five lectures that are widely publicized and nationally broadcast. Something of a British institution, the Reith lectures are often closely tied to the "issues of the day" and thus are often the occasion for sustained debate.

with what sort of thing information is. To ask "how much information do people want?" is just asking how long is a piece of string. But what they probably do want is *manageable* information and for that I believe it probably has to be what I call *extendable* information. Of course, a web site would be quite a good image of extendable information. But also, a preliminary conversation, a frequently asked questions sheet, and the information behind that. So, I think of it as *layered* information, that makes it more manageable, and when you begin to get the geography of it you can say "but tell me more about this." And, clearly in something complex like this, people do have opportunity to talk about it a lot, so they have the possibility of checking out their understanding, and checking out their misunderstanding, and checking out the limits of what's known and the limits of what they may expect. (Interview at Newnham College, Cambridge, 11 June 2002)

This definition of information as "layered" and "extendable" also corresponds to what O'Neill describes as the importance of risk information being accessible and *assess*able. The question of "how much information" can be recast as "what kind of information," "from who," "in what form" or "in what kind of information-sharing context." Time and deliberation are both key components to the model of trust through interaction O'Neill describes in what has been described as a Rawlsian approach to justice, which acknowledges the importance of both political and social inequality (Sayers 2001).

In her fourth Reith lecture concerning "trust and transparency," O'Neill argued that although access to information is crucial to accountability, *more information does not necessarily create more trust.* Indeed, she claims, "some sorts of openness and transparency may be bad for trust," and "more information may lead to uncertainty" (http://www.bbc.co.uk/radio4/reith2002/lecture4.shtml). The real enemy of trust, in O'Neill's view, is not the inherent uncertainty of an innovative technique such as PGD, the new genetics, or the modern world for that matter. It is, rather, deception: "Deception is the real enemy of trust," she writes. Fear of deception, or even an accustomedness to the ordinary euphemistic deceits of professional and political language (otherwise known as "spin"), may have caused expert testimony itself to appear partial and untrustworthy. In O'Neill's view:

> *Well-placed trust grows out of active enquiry rather than blind acceptance.* In traditional relations of trust, active enquiry was usually extended over time by talking and asking questions, by listening and seeing how well claims to know and undertakings to act held up. . . . Where we can check the information we receive, and when we can go

203

back to those who put it into circulation, we may gain confidence about placing or refusing trust. (O'Neill 2002b, 76, original emphasis)

This interactive model of trust as *relational* emphasizes the interrogatory, communicative, and deliberative aspects of accountability, and depicts it as something that is cumulative—*built up over time*. This kind of accountability is primarily based on communication, "talking and asking questions," and retrospection, "seeing how well claims to know and undertakings to act held up." These factors are clearly evident in the context of PGD, where trust is established not so much through the ability to ferret out deceit as via the accumulation of evidence *that there is no intention to deceive*. Indeed it could be argued that the lengthy conversations followed by the ability to reevaluate the PGD team's promises-to-delivery ratio offer an excellent example of how this kind of accountability accumulates. In contrast to the "one-way" model of scientific expertise as the neutral and objective truth, the transmission of which is by definition singular and definitive, O'Neill's model, like that of PGD practitioners, emphasizes time, repetition, accumulation, dialogue, multiple sources of information, and deliberation as the route to best practice.

This model of *how accountability accumulates* in the highly uncertain context of PGD could be extended to any hierarchical encounter involving the transmission of expert information. The verb "to inform" is defined as the ability "to give form to," which is in turn linked to the definition of "information" as *shaping or molding through instruction*. From this point of view of the power "to inform," it could be argued that the attempt to provide "neutral information" is always in tension with the inherent political differences between those who possess (and are in a position to give) information and those who are its recipients. For all these reasons, O'Neill's principle that "well-placed trust grows out of active enquiry" can be extended to suggest that accountability is fostered by a context in which "talking and asking questions" is encouraged, and in which the time required for this to take place regularly is acknowledged and provided for.

If it is thus the *context* of information transmission that significantly determines both its accessibility and its quality—as this study suggests—then it also follows that what makes information "manageable" is in part determined by *how much deliberation occasions its transmission*. "Good" information in this model must be dialogical, open, and interactive rather than one-way, unaccountable, or "closed." Quality information will take time to transmit, may need to be repeated, and will be available in a variety of forms. These must be subject to verification so they can be "actively checked." In O'Neill's model—in which information must be open to interrogation in order to be trustworthy—time is an important element, and is part of what enables trust to be *built* rather than merely assumed.

The place of sustained, critical, open questioning in such an equation thus becomes empowering and a sign of trust, rather than being unsettling or indicative of lack of trust. Ambivalence also acquires a more positive connotation, as uncertainty and equivocation are means to explore the multiple possibilities out of which a "best course forward" will be chosen. Questions thus become means of creating knowledge and trust, which, in the context of PGD, appear to be mutually enabling for both the PGD team and PGD patients. By interrogating patients about their reasons for pursuing PGD, clinicians set a tone of *protective skepticism*, and by expressing their own reservations and ambivalence about PGD they both qualify *and enhance* their own expertise. Both actions open the way for reciprocal questions, and as patients frequently confirmed, this process of foregrounding limitations and uncertainties enhanced their trust in the PGD team, particularly over time.

In the interview extract reproduced as the epigraph to this chapter (see the first page of this chapter), PGD patients Tony and Melissa express exactly this view. As Tony and Melissa emphasize, a positive sense of involvement is produced by thorough and lengthy explanations of failure. The experience of not being spoken "down to" prompts a comment that the PGD team do not seem like doctors at all. The form of this comment, as a question, elicits an affirmative negative: "no" (meaning "yes, I agree, they don't"). This is quickly followed by a positive qualification: "it's not because they're not professional"—a double negative that is followed by "or nothing." These layered, "knowing" comments dialogically evoke the stereotypical patronizing medical authority figure and the polarized, "one-way" view of the medical expert (who has knowledge) lording it over the ignorant patient (who has none) as the background to the comments that "they're really, really good to you," "they never ever speak down to you," "they go through everything," and "they're all really good."[6]

Such comments, while clearly appreciative, denote an exceptionalism that suggests many patients were not accustomed to such positive feelings toward medical and scientific experts. Moreover, the dynamic of patients feeling buoyed along by a sense of shared enthusiasm for treatment, along with their pride in being at the "cutting edge" of modern medicine, and their relief to be "on their way" once again, often after

---

[6] Although Tony's description refers to his initial encounter, and thus his first (rather than accumulated) impressions of the PGD team, his question to Melissa, "I don't look at them as doctors. Do you?" refers to the present, and to the sustained impression of trustworthiness and accountability that is affirmed by his subsequent comments.

years of feeling lost, stymied, or at the end of the road, is a noticeable theme in the patient interviews. As Terry said:

> It seems so unusual, people in a hospital being so nice to you. The people are all lovely. Jenny's fantastic. Peter Braude's very nice as well!

And as Tony commented similarly to Melissa:

> TONY: Even the embryologists are really nice, aren't they?
> MELISSA: Yeah.
> TONY: They show you round there, they show you the eggs and show you everything.

Often these expressions of appreciation were explicitly made in contrast to comments about previous medical encounters, such as Elizabeth's description of having felt in the past more like a statistic than a person:

> It was brilliant! It was *so* different. I mean, the way they treat you. They treat you like a person. When I was at the other hospital I felt like I was a statistic. St Thomas' was completely different. You always see the same people. . . . And they're all so friendly and that just makes so much difference, cause it's such a horrible process anyway, that when you actually feel like somebody cares about you (and I really do think they do) then it makes it a lot easier.

Elizabeth's emphasis on the St. Thomas' team's friendliness and her comment that "you actually feel like somebody cares about you" are in explicit contrast to her statement about the "other hospital." This level of care "just makes so much difference"—a comment she repeats three times. Noticeably, she also repeats for emphasis that "I really do think they do" (care), indicating that it is not only the performance of caring but a sense that this is genuine that "makes it a lot easier."

Like Elizabeth, Belinda also felt that her care was exceptional:

> They are very compassionate. I mean that comes across very clear and each couple is treated as individuals, as opposed to a case number.

She added that the Guy's and St. Thomas' PGD team take pride in their work, and in setting high standards:

> Yeah. And they're very proud of their results. You can see that they take pride in how well they're doing it, which is great.

For Nicholas, as for Tony, the team's confidence was further confirmed by their willingness to take the time to explain in detail what they do, their enthusiasm for their work, and their palpable desire to succeed:

But the guy—is he the embryologist? I mean the guy who wears the tweed suit. They're all great! You know, they get you and they sit you down, and you just—they're so desperate for you to get pregnant! And they're so into their work, that it's, it's, oh it's fantastic!

Not surprisingly, all these features of the PGD team's dedication increased patients' confidence in the team members' abilities, and trust in their professional opinions—including their belief in the team's unwillingness simply to "pamper them" by watering down unpalatable truths, as Elizabeth describes in a passage that is typical of many interviewee responses:

I think they're very honest. If they think that there's no point in us going on, then they'll say it. And I know that if they say it, then there really isn't. I trust them. I trust their professional opinions, and I know that when that happens, then that will be it. . . . I know that if they don't think there's any point going on they won't let us go through another just because we wanted to or something, you know? You don't need somebody pampering to you and telling you what you want to hear. You need to hear the truth. And they do that, but they do it in a sensitive way, you know? They are just really nice people. Really good at their jobs.

As is evident from Elizabeth's comments, some PGD patients value being treated with care and respect all the more *because it is not what they are expecting*, or what they have experienced in the past. This sense of feeling "special" is intensified by the unique and cutting-edge nature of PGD, and indeed even its likelihood of failure. As in any highly technically complex and arduous life-and-death medical care, there are high stakes for everyone involved, and the sense of working together to attempt to "beat the odds" creates a sense of teamwork that overrides some of the more traditional aspects of the doctor-patient hierarchy. There is, too, the sense that if they prevail, and a child is born, it will in an important sense be the progeny of a spectacular collective effort, and a literal embodiment of a dramatic medical achievement.

Inevitably, then, the sense of being buoyed along by treatment has many different sources and, while flattering to the practitioners, also has to be looked at critically. For example, it has to be asked, and patients ask themselves, as clinicians also ask each other, if there are ways that PGD in a sense "stands in for" the very outcomes it is intended to produce? Indeed, it could be asked if part of the dynamic that gives value to uncertainty is the place of hope, and the powerful role it plays in PGD (as it does also in IVF—which is also a form of achieved conception often pursued as a "last resort"). As Fiona Robson, one of the genetic counselors at Leeds, observes:

Some people just need to *pursue* it and then fail and, and then they can start again with something else, they just have to get that part out of their systems, as it were, and try. And we really *really* appreciate that you know, that we, would be supportive enough—I find it very difficult.

In her description of this "very difficult" situation, Fiona expresses sympathy for a couple's need to pursue PGD, almost no matter how remote their odds of success, in order to "get that part out of their systems" and to "start again with something else." This acknowledgment of the need to pursue PGD as a *goal in itself* is one of many indications encountered throughout this study of the complicated senses of obligation—to oneself, to others, to one's partner, and above all to existing, lost, or potential offspring — that drive PGD forward. In this case, the sense of an *obligation to try* may be seen by a genetic counselor as deserving of support and sympathy *independently of the chances of success*. For Fiona, the conflicting obligations she experiences as a PGD professional—of how to balance her desire to support couples by allowing them to try against her desire to protect them from the costs of doing so—are complicated by the question of who treatment is for. If PGD in some cases is not only a "hope technology" but also a "grief technology," what are the professional obligations that ensue? And, even more complicated, if undergoing PGD is for some couples (as some of the above extracts suggest) an opportunity to rebuild a trusting relationship toward their own reproductive potential, including its "failures," in the context of a caring relationship with top-notch medical professionals, is this trust itself a therapeutic outcome?[7]

The constant necessity for PGD staff to respond to (and to discriminate between) patients' needs to try, often to keep trying, and sometimes to try beyond the point of reason stems from the importance of PGD *as an end in itself*, described by patients in the previous chapter. This important therapeutic function of PGD, as *a form of trying* that can clear the way to move on to other things, is a recognized and valued outcome for a subset of PGD couples, much as it is also seen to be in tension with the possibility of such couples becoming stuck or lost in treatment, unable to bring themselves to stop.

Consequently, among the many aspects of PGD that clinicians find difficult, and worrying, is that they feel they need to offer a degree of guidance to couples about their chances of failure or success yet they

---

[7] In her comprehensive book on IVF failure, *When IVF Fails* (2004), feminist sociologist Karen Throsby argues that since failure is in fact the majority experience of IVF, by a ratio of four to one, there are many aspects of providing IVF that could be improved to take account of the experiences of the (often invisible) candidates who were unsuccessful.

cannot, except in rare circumstances, make a decision themselves. This, then, is the part of PGD that is difficult for "everyone": that it poses such difficult questions, to which there are rarely clear-cut answers, about which decisions must nonetheless be made, sometimes with very little time.

PGD thus exemplifies the increasingly common situation, inextricably tied to the power of modern medical technology, in which is impossible to know which answers or decisions are "right," and best practice must be based on *the quality of the decision-making process*, which in turn relies upon its perceived trustworthiness, or accountability. As O'Neill is quite right to point out, such a view inevitably shifts attention away from abstract bioethical principles such as protection of individual autonomy, and away from traditional bioethical concepts such as informed consent, toward the practical and logistical features of creating the trust that is necessary for patients to make active and genuine choices (2002b). Such a view, in its turn away from the "principalist" heritage of bioethics, closely matches the "sociological thinking" of Warnock, another "pragmatic philosopher" who favored principled, but also workable, solutions. By changing her question from "What is right?" in the abstract to "What will enough people think is all right?" in the historical present, Warnock sought to define workable grounds for accountability that remain open to, and are indeed strengthened by, a context of robust public interrogation. Indeed that is precisely how the HFEA has tried to keep "in step" with public opinion—by soliciting as many conflicting versions of it as possible. From this perspective, it is useful to be worried about PGD and to air one's concerns, and it is important for there to be ongoing conflict, as well as dedicated time for it to be heard and recognized.

Goal or Prison

Warnock argued that the one thing "everyone" could agree upon about the future of reprogenetic medicine was that it must be subject to limits of some kind—and, at a societal level, few would disagree. Significantly, however, the same cannot be said at a personal or individual level, where the opposite principle often holds—namely that nothing is impossible, everything must be tried, no stone can be left unturned, because a moral obligation to push beyond the limits is the only guide. This common, structural, and obvious tension between near-unanimous recognition of a social, legal, political, economic, and moral need for clear and established limits to technological "assistance" to human reproduction and heredity, and the desires of individuals in extreme circumstances to

209

break, defy, or transcend these same boundaries, is one of the foundational features of the history, politics, and sociality of IVF, embryo research, and PGD, and likewise of cloning, stem cells, and regenerative medicine.

Like prenatal diagnosis, PGD is a difficult context for lines to be drawn, and this is perhaps nowhere more evident than in the troubled relationship to hope it both furthers and relies upon. The reason the desire to try may itself be therapeutic, even if the treatment fails, is because keeping hope alive may be paramount. This is another part of the "difficulty" Fiona describes of enabling couples to express their desire to try everything, in order to feel no stone was unturned and they did the best they could—in order that they can be assisted to move on. Indeed, to "take all the hope away" by refusing a couple treatment sometimes, in Fiona's view, can seem cruel and unfair:

> Sometimes I feel like coming out and saying "this is cruel, this is not fair." This is not, you know, you know, people come with a bit of hope, you can't take all the hope away.

At the same time Fiona seeks to protect the last "bit of hope" that is left for some couples, she also feels a need to protect them against the tragic aftermath of PGD's probable failure. How can she know when it is best to allow couples to hold onto their hopes for as long as possible, or when they need some help letting go? She feels torn in her duty of care between protecting people against the "awful" experiences of anxiety and repeated disappointment and imposing the "cruel" foreclosure of hope—an ambivalence she describes as "very difficult." For Fiona, the full meaning of "autonomy" is only ever fully realized in the effort to recognize a couple's right to choose what they think is best for them, and to avoid being either overly directive or "paternalistic."

> As with a lot of things, I am just *amazed* how people deal with it, actually. So there's always that other side of it as well. You try to protect people from disappointment and a *huge* amount of anxiety and failure, and then, but you, of course you *can't*. That's got to be counterbalanced by people's autonomy and their *right* to go through an awful process if that's what they need to do. And you've got to allow—you know, so you can't be paternalistic either. And it's *very* difficult, very difficult.

Fiona's reference to the "other side" of her feelings about couples' need to try PGD—that she needs to protect them against the disappointment and anxiety of treatment, but also to protect their right to "go through an awful process if that's what they need to do"—references the signature ambivalence of PGD practitioners that is also a problem for "society" when faced with couples whose desire to "do anything" turns existing limits into seemingly arbitrary bureaucratic restraints. Even in a clinical context dedicated to individual care it can be difficult to know when enough is enough. The feeling of being "amazed" by what people are capable of dealing with, and by how well they can cope with almost unimaginably difficult and painful situations, sits uneasily alongside the "other side" of PGD, when a couple loses their sense of purpose and direction and feels defeated by the technique. While it is possible to leave PGD with more than you started with even if the procedure fails, because trying was enough in itself, the reverse is also true—that PGD can leave you with less than you had to begin with, resulting in a deprivation all the more painful because of how little there was left still to lose. The difference between PGD as an inspiring goal in the future—no matter how remote—and a prison of regret and despondency from which there is no escape is as impossible to anticipate as it is stark in its extreme polarity.

An important question that can be added to this difficult aspect of PGD treatment concerns the sense of obligation and responsibility patients feel—to themselves, to their partners, to their families, and in relation to social expectations—to "continue trying." As this study has shown, patients at the outset of treatment regularly describe the crucial difference between PGD as a step along the road and as the only road, although, in purely statistical terms, they know the latter is unlikely. As we have seen, there are important differences, however, between patients at the outset of treatment and toward its end point. For example, there are many aspects of treatment that are very positive regardless of the outcome, such as patients' high esteem for the clinical team, their sense of being on the cutting edge of medical technology, and the satisfaction of actively doing something and having a sense of direction—all of which create a sense of being buoyed along by the momentum of treatment, which fulfills the goal of "keeping hope alive." An important set of questions thus emerges about the ways in which a desire for children both reinforces, and is reinforced by, the experience of undergoing treatment.

Which Roulette?

The possibility of "losing everything" is one of the ubiquitous references to lotteries, gambling, "odds," "jackpots," and "chances" that permeate the language of PGD. The stark opposition between "losing everything" and "winning the lottery" defines the extreme stakes of PGD—the most sophisticated form of reproductive roulette in which the ultimate prize is

211

life itself. The temporality of PGD—already distinguished by its abrupt changes of pace—is further characterized, like the timeless space of hope, by its open-ended unpredictability. Heading into a cycle, be it their first or their fifth, a couple is entering into a period of a few hopeful weeks that will determine whether they will be able to create a successful pregnancy or not. If they succeed, all their effort will have paid off in the most spectacular fashion possible, and this is the reason PGD exists, because in roughly 25% of cases this is what happens.

These are not, on the face of it, terrible odds. Moreover, for those who succeed at PGD no price is too high, and however high is likely to be quickly forgotten if it succeeds. In contrast, the negative experience of failure affects a far larger number of couples, and is all they will have to take home. It is the dilemma of not being able to know at the outset of PGD which couples will eventually end up on which side of the enormous gap separating winners and losers that makes it necessary to prepare everyone for failure—a practice which, in the context of PGD, is seen to be protective and enabling for (and by) both patients and the professionals who work with them.

Since another distinctive feature of PGD, like parenthood, is the impossibility of sufficient "preparation," no matter how thorough or comprehensive the counseling or consultation process may be, it is also impossible to tell which couples will prove capable of coping with its demands, and which will break down under its pressures and abandon treatment, having failed to complete a single cycle. This dynamic produces the "very difficult" dilemmas described by Fiona, whereby a desire to protect patients against unnecessary additional stress and trauma must be "counterbalanced" against the patients' autonomy in reaching a decision that is right for them. Hence, while trying to avoid "paternalistic" behavior, it is also necessary to test patients' resolve and probe their understandings of what the technique can offer.

Hope Management

Preparedness for failure requires an education in hope management that begins with a process of hope adjustment, in the form of the "what PGD cannot do" phase of consultation. Approximately 50% of prospective patients leave at this stage, having decided PGD is not for them. The reasons for such a substantial rejection of PGD at this stage are not known, and comprise a vital but underresearched topic that awaits concerted scholarly attention. Among those who persevere, a second stage of hope balancing must be entered into, which could be called hoping and coping, because enough hope to cope with the demands of PGD must be maintained and nurtured, while too much hope (over investment) can compromise a couple's ability to navigate successfully some of the inevitable dips and drawbacks that beset even successful treatment cycles. Hope balancing, while essential, is euphemistic, as hope is neither fully amenable to control nor a quantity that can be easily reckoned in terms of less or more. The defining scale of hope is, after all, any or none. By definition, in a condition of uncertainty, any hope may be all you need, and there is nothing worse than none.

"Hope" was originally used in English to refer to a small enclosed area in the midst of wasteland. Its early referents also include a bay or an inlet where shelter may be found. Later definitions of hope refer not to a place affording shelter but a direction, or aspiration, as in "promise for the future" or "desire combined with expectation," which are both eighteenth-century definitions of hope. In Biblical depictions, hope, faith, and trust are closely aligned and refer to conditions of being: to live in hope or to trust in hope are the same as to have faith. Hope and belief are also linked through "the power of positive thinking"—a contemporary adage that depicts hope manifested as intention and merges this sort of "thinking" with its goal or object. To hope against hope (after Rom. 4:18) is "to hope where there are no reasonable grounds for doing so" or "to hope very much" (OED).

As in IVF, the importance of hope in the context of PGD is elaborate, ubiquitous, and prominent. At one level, PGD is all about hope—last hopes, only hopes, hope for the future, keeping hope alive, hoping against hope, and yet trying not to hope "too much." All the reproductive technologies are "hope technologies" (Franklin 1997) in their orientation toward future progeny, scientific progress, and triumph over adversity. Inevitably an aura of adventure and romance attaches to the quest to succeed in the pursuit of assisted conception, uniting PGD patients and the PGD team in a shared journey toward an elusive goal. For the PGD team, the hope that the technique offers is affirmed every time a child is born free from disabling genetic disease, or to a couple who could not conceive otherwise. For PGD patients, in contrast, the outcome is always singularly, intimately, and irrevocably theirs.

Economies of Hope

PGD patients and clinicians consequently occupy different economies of hope, since hope can "run out" for a couple but not for a clinic (or the science of PGD more broadly). While ample evidence suggests that both PGD and IVF patients whose hopes are not fulfilled by successful treatment remain hopeful toward the technique's prospects in the future, and toward its beneficial potential for other patients like themselves, there is no escaping the singularity of treatment outcomes for each individual couple, nor their inevitable primacy. Couples in this study, as in others, also say they feel they have contributed to hope for future improvements in assisted conception technology through their willingness to undergo treatment, which will have added to the cumulative experience on which clinical expertise is honed. Yet another indication of how couples who have had their own personal hopes disappointed continue to express hopefulness toward the future of assisted conception technology, and scientific progress more broadly, is the desire of many couples to donate their unused embryos to research.

Hope and failure are thus important places from which *to take account* of PGD, not only because they play so prominent a role in the lives of those closest to it, but because of the overemphasis on success that characterizes much media coverage of new reproductive and genetic technologies. It is this overemphasis on PGD as a "magic bullet" that is understandably worrying to clinicians and genetic counselors involved in the PGD consultation process, as it is not a technique that can be entered into with the expectation of having found a cure-all. As illustrated earlier, the "talking you out of it" stage of PGD consultation is a direct, preventive, and protective response to the costs of overestimating its capacities.

A related concern is the costs of unsuccessful IVF and PGD for the majority of patients who do not succeed. At one level, as Fiona argues, their right to fail is an expression of their autonomy. Indeed, as she points out, it is the right to try that many couples are seeking *as a goal in itself*. For some people, failure can bring rewards, such as having "gotten it out of their system," a sense of having tried everything, or a sense of having participated in and contributed to a pioneering area of medical scientific innovation.

On the other hand, the costs of failed PGD are by definition almost impossible to assess, although work has begun to be undertaken both into PGD failure and PGD refusal or dropout. More than any other area of reproductive biomedicine, PGD has been defined as a technique that is suitable only for exceptional patients. In both of the clinics involved in this study, a rigorous counseling and consultation process was used to ensure that patients did not undertake PGD with unrealistically high expectations of what it could achieve.

The cautious tendency toward PGD patient recruitment described in this study faces obstacles as PGD expands. The increasing tendency for PGD to be used to improve IVF success rates through aneuploidy screening will increase the overlap between PGD and IVF, making PGD both a fertility enhancement technology and a branch of clinical genetics, and blurring the boundaries between preimplantation genetic diagnosis and screening.

The introduction of multiplex genetic analysis would potentially complicated the diagnosis of single-gene disorders by offering additional simultaneous diagnostic possibilities. Most prominent in the period 2001–4 is the expansion of PGD to include HLA-type tissue matching, as well as genetic diagnosis, for couples seeking a "savior sibling" to donate cord blood or bone marrow to an existing child suffering from incurable genetic or metabolic disease (as discussed in chapter 1).

The Unaccountability of Hope

One of the major problems resulting from the importance of hope to technological improvement, and particularly to new reproductive and genetic technologies, is its *unaccountability*. Hope is not easily explained, nor readily interrogated. It is neither accessible nor assessable, and it is frequently unmanageable. Indeed, as both a noun and a verb, hope eludes easy classification. It does its work primarily as a condition.

The condition of hope has been the subject of extensive disagreement between those who see it as enabling and those who see it as disabling. Marx famously described religious hope as an opiate that led people to accept their own exploitation. This pessimistic view of hope is similar to that recounted in the tale of Pandora's box, from which disease, pestilence, and other ills escaped while hope, the worst of evils, remained behind to ensure that human misery would be borne out in lifelong suffering. Nietzsche shared this view of hope as disabling in his denouncement of it as "the worst of evils" because it "protracts the torment of man" (*Genealogy of Morals*).

As anthropologist Ghassan Hage observes, "To think about hope is exceptionally frustrating, in that it can sometimes seem as if one is examining something as vague as 'life,' given the multitude of meanings and significations associated with it" (2002, 9–10). Hage's work introduces the concept of *social hope* to address the problem of racism in society. Borrowing from the work of psychoanalyst Anna Potamianou (1997), who distinguishes between "hope for life" (activity, investment) and "hope against life" (passivity, withdrawal), Hage emphasizes the importance of discriminating between different kinds of hope, and *the alignment of individual and social hope* (see also Brown 2003; Brown and Michael 2002, 2003).

Such a view has numerous implications for the kinds of hope at work in the context of new reproductive and genetic technologies, as well as introducing a means of thinking about hope in terms of accountability. For example, Potamianou's distinction between "hope for life" and "hope against life" takes on a significant set of added connotations in relation to IVF

215

and PGD. Hope for new life, referring both to offspring and to the life they will enable, strongly defines the context of PGD. The danger of "hope against life" is one of the ways "too much hope" can become destructive, sapping a couple of their will, or sense of control over events. The impossibility of defining a clear path between these two forms of hope, and the significant role of chance in determining reproductive outcomes, is one of the defining conditions of reproductive and genetic biomedicine.

It is nonetheless apparent in the context of PGD, however, that although both the necessity of hope and its dangers are "impossibly" copresent and intermingled, this is not an entirely "unaccountable" experience for either patients or PGD professionals. For example, while the difficulty of finding an end point to treatment, and abandoning hope, may be impossible to "answer," it is not impossible to reckon, to enumerate, or even to reason through, if not to give a reason for. To be accountable is not necessarily to be fully explicable; it also refers to being called to account, or to being interrogated. To be called to account may require the presentation of evidence without presupposing a judgment. For the PGD patients in this study who valued uncertainty *as a form of accountability*, the withholding of clinical judgment was crucial, in order for them to be *enabled to manage their own uncertainty, rather than have it be managed by others*.

Conclusion

The "accounting" offered in this chapter, then, is not so much a *search for answers* as a *search for questions*, and in particular an effort to document the questions that are asked of PGD by those who are closest to it. Following Onora O'Neill's model of accountability as being "open to question," this chapter suggests that the way PGD, an innovative and "uncertain" technology, is made accountable is in precisely this manner. From the outset, prospective PGD couples are asked to give a robust account of their desire to pursue PGD, and clinicians provide equally detailed accounts of its benefits and drawbacks. This chapter suggests that this questioning process includes, *and is strengthened by*, the many difficult and unanswerable questions PGD patients and clinicians ask themselves about the costs and benefits of the technique—a point that is further illustrated by the complexity of the questions and the difficulties that are articulated both by patients and by professionals (which, as noted at the outset of this chapter, contrast markedly with the oversimplification of these issues in many accounts of the "designer baby" question at a further remove).

Contrary to the view of PGD as careering forward out of control, ironically, by offering "too much choice," the evidence presented here suggests a much more careful and thoughtful engagement with the difficulties presented by PGD, for example in the range of ways couples protectively plan ahead how they will move on if treatment is unsuccessful, the difficulties they acknowledge in knowing where to "draw the line," and the ambivalence expressed by professionals about the decision to encourage or dissuade borderline PGD candidates. While PGD would-be parents are rightly described as determined, it is their ambivalence that allows them to move *simultaneously forward and back and forth* between different alternatives to PGD.

Turning to the question of how accountability within "the topsy-turvy world of PGD" is relevant to the wider questions posed about the future of reproductive biomedicine in the context of its rapid expansion, a few speculative conclusions can be drawn. Will there be increasing pressure on couples to undergo procedures such as IVF and PGD in order to be seen to be doing "everything"? Is there too much emphasis on having children by these means instead of others? How should the expansion of PGD be regulated in the future?

If it is "best practice" in the clinic for PGD to be the subject of ongoing, critical questioning, how can this be translated into "best practice" in society for regulating future uses of PGD? How can the individual hopes of patients, often organized as a desire to "do anything," be aligned with the social hope for techniques such as PGD (i.e., that such techniques will be permitted subject to strict regulation, and that their use will be legitimately authorized only if legally enforced limitations clearly define their sphere of appropriate use)?

In the following chapter, the alignments and fractures between individual and social hopes for PGD, and the problems of accountability they engender, are explored in the context of reviewing the overall conclusions of this book. Building on the arguments offered in this chapter—about the prominence of hope, uncertainty, and ambivalence in the context of PGD, and their relationship to its "accountability"— the conclusion offers a social model of PGD, arguing that PGD is in many ways the opposite of the dangerous and dystopian "designer baby" characterization, which so unhelpfully continues to dominate much public and media discussion of the increasingly difficult choices and challenges PGD presents to individuals, couples, professionals, and the wider society of which they are a part.

Conclusion
PGD Futures?

As noted at the outset of this account of PGD, there are several reasons why it has become not only a focal point of contestation over the future of reproductive biomedicine, but itself a condensed signifier of broader anxieties symbolized by the figure of the "designer baby." Situated at the intersection of reproductive and genetic technology, PGD inevitably poses difficult questions about the increasing role of genetic choice in the context of reproductive intervention and the ability to "assist" both conception and heredity. It would be worrying if such technology did not elicit concerns. Nor are many of these concerns neatly classifiable into preexisting categories. This is the main reason why the effort to collect and analyze public, professional, and personal perspectives on PGD from a range of sites and locations is an obvious, if somewhat awkward, task. For it is here, embedded in the languages and practices already at work in the "topsy-turvy world of PGD" that we can find an existing sociology of reproductive and genetic technology "ready-made" and already to hand.

"Designer baby" is a broad, paradoxical, and complicated term that can be understood in a range of ways—as accurate or inaccurate, as celebratory or pejorative, as desirable or undesirable. Among the majority of the PGD patient interviewees, it was resented as a term of abuse, much as the term "test-tube baby" had been by a previous generation of IVF patients. As Kirsten, a PGD patient, protested:

> It's not as if we're going to this unit with a shopping list saying we want this and this. It's more a case of, we're looking for this [genetic disease] because that's what you need to eradicate!

In another case, however, a PGD couple joked about their "Gucci baby," and in a national newspaper column written about her experience of having a PGD baby, journalist Leah Wild declared:

> I am hoping to have a designer baby. Any offspring of mine will not be God's gift: I fully intend to exploit the latest medical advances to ensure my child is as perfect as possible. This involves selecting some

embryos and eliminating others. He or she will be chosen according to how good their genes are. (Wild 2000)

Wild's insistence on claiming her designer baby's selective origin, though it may appear celebratory, is aimed at counteracting the negative connotations of PGD, a technique she "would prefer not to have," and that is the result of her translocation:

> My choices, as those of the other "genetically disadvantaged,"
> are not the same as the vast majority of the reproductive population.
> I either make this genetic selection, or risk further multiple
> miscarriages and accept infertility. I would prefer not to have to.
> No one in their right mind would opt for the procedure of PGD instead
> of what is quaintly called "the old fashioned method." (Wild 2000)

For other couples, the term "designer baby" confused the eradication of genetic disease with the selection of desired traits such as hair and eye color. As Scott put it:

> I don't agree with people who want to choose blue-eyed, blonde hair
> babies, because that's what they want, but there's a hell of a lot of
> people out there that are doing that for other reasons and that's
> really, I guess, what this research is pitched at. It's not pitched at,
> *because they want to go down an aisle and pick a baby off the shelf*
> with blonde hair and blue eyes. They're doing it to eradicate genetic
> diseases!

Scott's comments demonstrate a strong distinction between "designer babies" and the effort to prevent unnecessary suffering, but also an awareness of the potential for the technique to be used in ways he disagrees with. Similarly, while Belinda and Nicholas could see the advantages of expanding the scope of PGD to prevent a wider range of diseases, they could also see the antisocial potential of increased genetic selection:

> BELINDA: Well, *certainly* I think, if you're going to all the trouble of
> putting the embryo at risk by taking a biopsy of it, for just *one*
> thing . . . it would be *fantastic* if at the same time you could screen
> out, um, [*pause*] Down's syndrome and spina bifida or any one of
> these . . . really important diseases that have a high incidence. If
> you could just do a general screen along with your one *particular*
> screen, that would be fabulous.
>
> NICHOLAS: Yeah, I think Belinda's right. I mean unfortunately this
> sort of scenario is *potentially* open to abuse. . . . I'm *massively*
> against it, I think if there is ever any attempt to start selecting the
> sex of your child at the outset, regardless of whether they have blue

219

eyes or dark hair, height, intelligence, whatever, I mean that should just be jumped all over.

As other couples, such as Anne and Daniel, emphasized, they were not even trying to eliminate disability but rather to prevent a painful, certain, and premature death.

> ANNE: I mean if we was to find out that we was carriers of something else, and we were just going to produce a child that would inevitably be *disabled*, or whatever, we wouldn't *use* PGD. We see it as something that can prevent children dying, basically, that's why we're using PGD.
>
> DANIEL: And it's not even a case of um, I think with SMA, its not even a case of "well they might die." They *will*! There's no question about it, they will!

As argued earlier, PGD patients frequently come to the technique with a significant history of reproductive loss and trauma behind them. They have often followed a lengthy and complicated path to its door, and have opted for its demanding regime out of a sense of obligation to avoid imposing harm on their potential offspring, or upon themselves, or both. As has also been emphasized, PGD is carefully presented to prospective patients deliberately to avoid the misperception of it as a "magic bullet." Instead, couples seeking PGD are rigorously questioned about the extent to which they have considered alternatives.

However, while this study has provided ample evidence to counter the stereotype of choosy parents seeking perfect offspring, the analysis presented here has also sought to avoid dismissing the fears associated with "designer babies" as irrational or misguided. To the contrary, data collected inside the "topsy-turvy world of PGD" consistently showed that the closer people were to PGD the more likely it was that they would have both strong emotions in response to it and significant concerns about its use, including ambivalence about its future uses. Indeed, evidence of substantial critical thought about PGD was both consistent and universal. Even when the reasons for pursuing PGD were "obvious" or "clear-cut," it was never presented as an easy option, or a straightforward one. To the contrary, everyone with experience of PGD was well aware of its many risks and drawbacks.

It could be argued PGD is primarily sought as an option in order to preserve a biological tie to one's "own" offspring, but this study did not produce evidence that the desire for biogenetically related offspring was a primary motivation to pursue PGD. Rather, it found that even couples who described it as their "only choice" had explored alternatives. A careful reading of patients' comments even belies a simple relationship between the desire for children and PGD, or between PGD and childlessness. As argued in chapter 5, PGD patients frequently had given consideration to a wide range of future options, in order deliberately to hedge their chances of succeeding with PGD by positioning it as one of several possible "roads." As indicated by previous studies of IVF, as well as Linda Layne's analysis of pregnancy loss, the experience of failure proved to be a significant component of the "topsy-turvy world of PGD," as was the persistent tension between the normative pressures on couples to have children and the 75% odds of the technique failing to produce a successful pregnancy (Franklin 1997; Layne 2003; Thompson 2005; Throsby 2004).

These dilemmas were accompanied by others that are familiar both to couples who have undergone, or who are undergoing, PGD and to the professional staff who work most closely with them. In addition to efforts to minimize the costs to patients whose treatment fails are concerns about the difficulty for some couples of finding an end point to treatment, unrealistic expectations of its success, lack of preparation for the number of technological obstacles to successful treatment, and the impossibility of preparing for the emotional and physical demands of the procedure.

A striking finding of this research was the very high level of appreciation PGD patients expressed toward the scientific and clinical staff—a finding that is clearly related to working with a sensitive and experienced professional team but is nonetheless also indicative of a pattern, or even principle, that would repay further study. Somewhat unexpectedly, this appreciation was rooted in a strong correlation between the PGD team's efforts to emphasize the drawbacks of treatment and couples' desire to undergo it. A significant finding, then, was that patients were much less concerned with the amount of information they received (too much or too little) than with its qualities, and in particular its context of delivery. It is also notable that one of the qualities of information patients appreciated most was the acknowledgment of uncertainty, ignorance, and failure on the part of the medical and scientific staff. This repeated finding can be characterized as "uncertainty value" to emphasize the paradoxical importance of uncertainty to what is often described as accountability or transparency. As Ian emphasized:

> I suppose people were so *honest* with us, you know, they, they're really only, only too happy to sort of, um, let us know the sort of limits of their knowledge and the sort of great extent of their ignorance really and what these things may mean. No one could actually specifically say what this inversion might result in, just a list

of possibilities (yeah). . . . [The doctor] was very positive and I think that's what you need at that early stage.

The fact that patients valued explanation, and in particular the PGD team's acknowledgment, as Ian says, of "limits of their knowledge," creates significant resonances with the arguments offered by Onora O'Neill on the questions of "trust and transparency" she raised in her nationally broadcast BBC Reith lectures in 2002, in the midst of considerable public debate over PGD.

The finding that patients value uncertainty can be extended to suggest that *patients would rather manage their own uncertainty than have it be managed by others.* This preference suggests that patients' decisions also act as expressions of reciprocal respect for clinical restraint, again demonstrating the importance of *limits* as a form of protection both of patients and PGD team members. The relationships between limits, trust, and accountability at the heart of determining how science and society can "keep in step" are not unrelated to these "microdynamics" of the clinic, which may, in fact, illuminate the basis of wider social practices more fully.

This study confirms the views of other ethnographers of new reproductive and genetic technologies, in particular those of Rayna Rapp (1999) and Jeanette Edwards (2000), who have demonstrated the depth of engagement and subtlety of "lay" understandings of genetics and genetic decision making. Similarly, it extends the arguments of both Margaret Lock (2001) and Paul Rabinow (1999) that understandings and models of life, vitality, and reproductive substance show marked national characteristics—in this case particularly with reference to the importance of *equity and altruism* observed in discussions of PGD. It further confirms the insight of Strathern's analysis of the "duality" or "hybridity" of kinship (2005), and the translation of this feature into what might be described as the "digital" capacity of even what is biogenetic, or supposedly fixed, about genealogy, to be readjusted and remade into its opposite, as Thompson also convincingly demonstrates (2005).

It was more difficult to find evidence either of what Lippman has termed "geneticisation" (1992) or of what Finkler (2000) has defined as the reconsolidation of genetic kinship through DNA. While there was ample evidence of the "gene talk" described by Duden and Samerski (2003), and of the "somatic individualisation" described by Novas and Rose (2000), this study, like those of Thompson (2005), Hayden (1995), Ragone (1994), and Konrad (2005a, 2005b), revealed considerable flexibility in terms of how genes, genetic ties, and genetic relationships were described and imagined. It could be said this study, although indicative

222

rather than representative, confirms, as Rapp (1999) and Edwards (2000) also do, that greater proximity to genetic information may be experienced as increasing its uncertainty *as well as* its fixity or determinism (see in particular Strathern 1999). The tendency to dichotomize these possibilities (either DNA "tells you who you are" or it doesn't) may thus be an analytical tendency at odds with a growing amount of empirical data (see esp. Lock 2005).

Two crucial areas in which little or no research appears to have been undertaken are why large numbers of people (in the London clinic nearly half) who are offered PGD do not continue, and what they do instead. Further investigation of both of these issues could potentially identify important dimensions of patient expectations about treatment, as would more data on how people conceptualize alternatives to both IVF and PGD. Data from such studies might also yield findings that would be useful in understanding the difficulties some couples experience in finding an end point to treatment—one of the most difficult issues raised by PGD—and the complicated interplay of what are described here as "technonormativity" (the pressure of normative expectations about the benefits of scientific and technological progress) and "bionormativity" (pressure to conform to patterns that are considered to be biologically normal).

This study identified a higher than expected overlap between infertility and genetic disease, somewhat contradicting the view of PGD as different from IVF because it is undertaken among a largely fertile population. This is one of several issues that could be explored under the broad umbrella of the changing relationship between assisted reproduction and assisted heredity, and may also be related to some of the questions raised above. For example, if, as Charis Thompson argues in her structural account of the "biomedical mode of reproduction," the effort to "rescue the lifecourse" can also lead to its rearrangement, and the enlistment of technological assistance in "family building" can alter what that comes to mean, it is clear the components of PGD have to be disaggregated into more discrete components (Thompson 2005, 266). How many PGD candidates are searching for confirmation of something linked to, but not altogether identical with, having a child? How is the "conjugal choreography" of couples' perceptions of how sexuality, reproduction, and technology are linked at issue in their pursuit of an increasing number of reproductive "alternatives," including PGD? What is the significance of the fact that PGD is not, like IVF, "imitating" the creation of life but is instead a form of selection aimed at identifying and eliminating the origins of serious genetic disease? Are these questions displaced by the provocative moniker "designer baby," and, if so how might this be remedied?

Whereas IVF was strongly naturalized from its inception through the idiom of "giving nature a helping hand," and indeed through the very term "assisted" conception, PGD is much more explicitly represented as a direct intervention into, and interruption of, the conceptive process. As noted in the preface, the whole point of PGD is precisely *not* to give nature a helping hand but to *prevent* nature from doing what it might do "naturally," by making sure certain possibilities can be eliminated in advance. It is in this sense that the term "design" is accurate, as a description of fulfilling a plan or intention. The "designer" baby in this sense is one that has been predetermined not to have a particular trait—the opposite of how this term is usually used.

If to be born and to be made are not, then, as some might imagine, "opposites" organized in a dichotomy akin to the natural and the artificial, or the social and the biological, but are, instead, *historically unified in the idea of "reproduction,"* then the questions raised by PGD, so often depicted in terms of a new dawn, a new era, or a qualitative change in "who we are," should be seen as familiar—literally. It may be that the speed with which IVF has come to be seen as "normal" derives from this same principle—that assistance to reproduction does not denaturalize it. PGD extends this principle into the realm of "assisting" heredity. According to this view it is the *nonnovelty of PGD*, however, that brings it into conflict with the perceived need for limits. In other words, and as the "savior sibling" and Diane Blood cases vividly illustrate, it is not what is "science-fiction" about PGD that pushes it into the spotlight, but rather what is perfectly familiar and understandable about how it works that positions it as a pivotal technique. It is pivotal because it is, at present, a "tipping point" between conflicting obligations: the obligation to remake conception, and the obligation for there to be limits to intervention into being born. Importantly, the capacity for PGD to embody conflicting obligations exists at every level of this study—from mainstream public debate, to couples' anxieties over treatment, to high-profile campaigns to extend the use of PGD, to the private worries of PGD practitioners.

As we have seen, PGD was the subject of considerable debate during the course of this study as it became linked to a number of controversial cases, in particular those of the Hashmis and the Whitakers. This debate, as discussed in chapter 1, can be described in terms of conflicting obligations as well as the issue of "single desperate cases," which create pressure for change. Importantly, they demonstrate the form of this conflict as well, in that the obligation to limit new genetic technologies will always be in conflict with the desire to improve them on behalf of people who suffer from genetic disease. This conflict will become most acute when lifesaving treatments are clinically possible, but legally banned.

The cases of the Hashmis and the Whitakers also demonstrate that even technologies that may appear "repugnant" to some will be desirable to parents of children who are dying. These "desperate" individual cases show that social efforts to limit technology will be weakest at the point where they conflict with the moral imperative to save lives in the presence of the technological capacity to do so. This conflict is partly driven by the view that a parent's desire to do everything possible on behalf of his or her offspring *cannot be limited*. It is also an indication that the period of debate prior to the clinically proven viability of a technology is in some respects qualitatively distinct from the period of debate following a technology's successful clinical debut (Franklin 2003d).

The task of the HFEA of "keeping in step" with public opinion, and determining what the public will be "comfortable" with, will thus continue to be open to the vicissitudes of public equivocation, which can be solved only through procedure and deliberation, because there is no way "finally" to determine what is "right" for everyone. As the previous chapters have shown, the HFEA seeks an "objective mix" of opinion, based on both directed and open consultation, closely resembling the "sociological thinking" pursued by Warnock, in order to perform the paradoxical task of both setting strict limits and changing them. The empiricist political philosophy of Onora O'Neill also favors an approach to justice and accountability through a strategy that recognizes social inequality.

A consequence too of the high-profile cases that created intensive debate about PGD and "designer babies" in Britain in the first years of the twenty-first century is that they created an opportunity to observe the ways in which reproductive genetic technologies such as PGD are both pushed forward *and held back*. This suggests that new genetic technologies are not "beyond" society: they are the subject of constant social negotiation. The cases described in this book, and the controversies and debates surrounding them, demonstrate the importance of social relationships to the formation of the hopes and desires associated with technological innovation. Hence, rather than being an example of technology racing "ahead" of society, PGD provides a useful example of how social desires often race far ahead of technology.

The theme of conflicting obligations is relevant not only to the "wider social issues" of PGD or its regulation, however. PGD is a "tipping point" for tensions, conflicts, and "impossible questions" at every level. The fact that couples have described PGD as "their only choice" cannot be taken literally, because, as was evident from several of the interviews discussed in chapter 5, this statement is not incompatible with the ability to describe several choices as equally preferable alternatives to one another, of which PGD is but one. The statement that PGD is "our only

225

choice" is rather a statement of obligation—it describes a sense of obligation to "try everything" that is by no means restricted to the context of assisted conception, and by no means unfamiliar as an expression of a desire to "do everything possible."

As we have also seen, it is precisely because PGD was *not* the "only choice" that clinicians and genetic counselors felt conflicting obligations toward patients at several levels—in terms of how much information to provide, how long to pursue treatment, and how many decisions to delegate to patients. The PGD couples who were interviewed for this study described conflicting obligations in terms of how long to pursue treatment at the expense of other aspects of their lives, and how much of a burden such treatment might impose on their existing children or families. They raised concerns about the health consequences of PGD for women partners who might have to undergo several cycles of ovarian stimulation, and about the effects of PGD on their relationships and their emotional and mental health.

Two episodes that Sarah Franklin recorded in her field notes toward the end of this study encapsulate this tension at the heart of PGD, which, in its form, repeats with striking similarity the social "ambivalence" toward this technique.

Feb 10 2003

I arrive in the doctor's office at clinic just past 2pm and Peter is already here, having just come in from a final consultation with a couple who have reached "the end of the road." Peter is visibly distraught. He sits down heavily at the desk and puts his head in his hands. "Sometimes I just don't know if we should be putting them through this," he says, staring at the floor. "And sometimes I just don't know how they can put themselves through it either," he adds, looking at me with a resigned expression that asks for no reply. His entire slumped body expresses the weight of his concern—a mixture of grief, frustration, disappointment, and ambivalence. I have no doubt he is suffering himself from this experience, and that he must wrestle with this painful question every time a couple under his care leaves his clinic possibly worse off than they were when they first met him. It is hard to imagine anything upsetting him more than the idea he has contributed to a couple's unhappiness. It seems to me he is acknowledging the inevitable "gray area" that makes it impossible for anyone to know for certain whether PGD is "right" or "wrong"—not even him. It is so amazing when it works, but so incredibly costly when it fails.

Feb 17 2003 (a week later)

Peter came in after a particularly difficult case and, in his usual
generous and helpful way, started to fill me in on the difficulties. These
were numerous and included one of the patients being a carrier for
Huntington's. Again, with his characteristic intensity of concentration,
he expressed his frustration: "I just wish we could *do* something for
these people! I just wish there were more we could offer them."
I cannot help but be struck by the similarity of Peter's emotional
reaction to the one he expressed only a week ago, in what might be
described as the "opposite" direction. Last week he was as sincere in
his doubt about whether PGD should be offered to anyone as he is
today in his conviction to improve it so more people can benefit. I
think this completely summarises the ambivalence of PGD: how is he
to know how best "to do no harm" or to provide relief of human
suffering, when the very technique itself can literally be either a life
saver or a life sentence?

These two extracts, succinctly exemplify the range of reactions to
PGD among both those closest to it *and those for whom it is an impersonal symbol of genetic medicine.* As Peter's reactions demonstrate, it is
perfectly possible to feel deeply ambivalent about PGD, while ultimately
feeling it is right to offer it as carefully as possible to as many people as
possible, and indeed to ever more groups of people for an ever-widening
range of conditions. This cannot be described as a dilemma produced by
the technology "itself," much as it often appears to have "a life of its
own," which, in a sense, it now does. PGD came into being out of the
same complex matrix of emotional desires—of wanting to help, wanting
to relieve suffering, wanting to be able to know and do more, and never
knowing "how far to go"—that characterize it today. Perhaps some clinicians, or scientists, or patients are unequivocal at the outset. But no
one working closely with PGD can feel completely confident it is 100%
"the right thing to do." Even people whose treatment succeeds, as we
saw in chapter 5, express caution about recommending PGD to others,
knowing it is so likely to fail. This is why even among its keenest supporters, attitudes toward PGD are mixed, ambivalent, and qualified.

Among the costs of PGD, as we have seen, are the possibility that its
very existence acts as a pressure on couples to attempt it, and this returns to the problem of choice in the context of PGD. Does technology
narrow choices by creating an imperative or a trajectory toward technological solutions? Or is it the case instead that prospective patients do
not have enough choice in spite of the technology, especially because it

so often fails? What other choices would need to be available for PGD to feel less like "the only choice"? How can the hopes of individuals that are pinned to the prospect of successful PGD be reconciled with wider social hopes and ambitions, in the sense that Ghassan Hage imagines as a more collective set of shared aspirations? Is this what the HFEA is doing by establishing the agreed-upon limits than enable "steady progress" to continue?

It is a ubiquitous and striking feature of "the topsy-turvy world of PGD" that it is intensely emotional. Traditional psychological and sociological models of emotions, like much popular discussion of them, are based on the assumption that they have an internal origin, or source, that is externalized, or expressed, in reaction to stimulus or provocation. However, more recent theories of emotions depict them as the product of *external contingencies*, or contact. Sara Ahmed, for example, suggests that "feelings" do not simply originate within an individual person, as if already there to be elicited, but rather are formed through patterns of interaction that channel our emotional responses into specific forms (2004). Another way to say this is that emotions are in a sense both socially shaped and even generic: grief, anger, happiness, fear, and hope are emotional formations that have the same degree of social specificity as forms of embodiment, and the same degree of imbrication within institutions as forms of labor, economic activity, or family formation.

Why should it be surprising that the intersection of reproduction, heredity, and scientific innovation will both take shape within and engender particular patterns of affect and hope? It should be perfectly obvious that, in the remaking of "the facts of life," identities and relationships are also being remade, as well as the values and logics they embody. As this book has shown, PGD is the context of considerable moral, political, ethical, philosophical, and sociological reasoning about what its practice involves, and how it should be both limited and expanded. In contrast to the spatialized idiom of science racing ahead into uncharted territory creating a new reproductive frontier, this book demonstrates that it is instead a complex deliberative process, and a highly restrained and cautious attitude, that gives the "topsy-turvy world of PGD" much of its texture, character, and depth—perhaps epitomized by its "topsy-turvy" career in the history of public and parliamentary debate of human reproduction and embryology in the United Kingdom.

So long as PGD remains a "tipping point" in debates about the future of reproductive and genetic redesign, and a focal point for the question of how to balance being born and being made, it will serve as a *technology of accountability*, as well as being a technology of hope, of grief, and of "reprogenetic assistance." As such, it will need to remain the object of a "hot-pot" of opinion, and be subject to verification and "extendable" interrogation. To the extent that this accountability is already embedded as a form of what might be described as sociological attention, so we are even more of our own making than we thought.

Appendix

PATIENT INFORMATION SHEET

Definitions of Genetic Knowledge and Pre-Implantation Genetic Diagnosis:
An Ethnographic Study

The introduction of new genetic technologies in the reproductive process raises many challenging social and ethical issues. From February 2001 to August 2002, a sociological inquiry into Pre-implantation Genetic Diagnosis (PGD) is being undertaken at Guy's & St Thomas's Hospital and Leeds General Infirmary to understand more about the meanings of genetic choice. Observations of intake interviews, medical procedures, team meetings, and other clinical activities will be supplemented by semi-structured interviewees with PGD patients from a range of backgrounds, with different medical histories, and at various stages of treatment.

The study is jointly funded by the Economic and Social Research Council and the Medical Research Council. One of its primary aims is to increase co-operation between medical professionals and social scientists in addressing the impact of the new genetics in its broadest sense. Outcomes from the project will help to improve clinical practice, to inform government policy, and to define the licensing and regulatory policies governing the application of genetic medicine.

Professor Sarah Franklin and Dr Celia Roberts from the Department of Sociology at Lancaster University will be working with the PGD teams at Guys & St Thomas's and the LGI on all aspects of the study, and will be conducting the interviews with PGD patients. These interviews can be arranged to take place either at your treating hospital or in your home. They could be with either partner, or both. Because of travel restrictions, interviews that are not held in or near the clinic will need to be in the Leeds, Lancaster or Manchester areas.

The interviews are designed to understand the experience of undergoing PGD from a patient's perspective. They will cover how you first learnt about PGD, the decision-making processes involved in using it, and what your experiences of it have been so far. Interviews will be open-ended and will be of 1-3 hours duration.

Both the interviews and consultation sessions for which prior approval for observation has been granted will be tape-recorded and transcribed for the purposes of research. *No record of any patient's personal details will at any point be made public from this study.* Although it may be useful to reproduce extracts from transcriptions from interviews and consultation sessions for the purpose of scholarly publication, *no identifying details will be revealed.* The very strictest levels of confidentiality will be maintained in accordance with the regulations of the Human Fertilisation and Embryology Authority (HFEA). Tapes will either be destroyed once the study is finished, or can be sent to you to keep if requested. Copies of interview transcripts will also be made available on request. Participation in the study is entirely voluntary and all interviews will be arranged at the interviewee's convenience.

For more information about this research, contact Professor Sarah Franklin, Department of Sociology, Lancaster University, Ph 01524 594187; fax 01524 594256; email s.franklin@lancaster.ac.uk, or visit the project webpage at: http://www.comp.lancs.ac.uk/sociology/IHT/Mainpage.htm.

Figure A.1. A Patient Information Sheet is required for ethical approval of the study and must provide a clear and accessible account of the research being conducted and its implications for participants.

CONSENT FORM

Project Title: Definitions of Genetic Knowledge and Pre-Implantation
Genetic Diagnosis: An Ethnographic Study

St Thomas' Hospital Research Ethics Committee Code No: EC01/048

Researchers

Prof Sarah Franklin	Sociology Department, Lancaster University
Dr Celia Roberts	Sociology Department, Lancaster University
Prof Peter Braude	Guy's and St Thomas' Hospital, London
Mr Anthony Rutherford	Leeds General Infirmary

This study aims to describe and analyse the complex social and ethical issues raised by pre-implantation genetic diagnosis (PGD). The study is funded through the Economic and Social Research Council and Medical Research Council and is a joint undertaking of researchers at the Department of Sociology, Lancaster University and the Assisted Conception Units of Guy's and St Thomas' Hospital (London) and Leeds General Infirmary.

Interviews and taped consultation sessions with people involved with PGD treatment form an important part of this study. Interviews will be of 1-3 hrs duration and will be audio taped and transcribed. Consultation sessions will not be recorded without prior consent. Copies of the transcripts will be made available to interviewees at their request. At the end of the study, all tapes will either be returned to interviewees or destroyed.

I (name) _____

of (address) _____ state that

- I have read and understood the patient information sheet accompanying this consent form.

- I understand that this consultation/interview will be audio taped and transcribed for the purposes of research. Some parts of the transcript may be quoted for the purposes of academic publication.

- I understand that I will remain anonymous at all times. My name will not be attached to either the tape or the transcript, and will not be used in any publication.

- I understand that I may terminate an interview, or cease to have a consultation session recorded, at any time. This will not have any consequences for the care I receive as a patient or for my relationship with any hospital staff.

- I agree to take part in this recorded consultation/interview.

Interviewee	**Witness**
Signature _____	Signature _____
Date _____	Date _____

FIGURE A.2. The explanation of the study, and the process of signing a Consent Form at the outset of an interview, are important in terms of ethical best practice, and also "frame" the interview as a formal exchange.

References

Ahmed, Sara. 2004. *The Cultural Politics of Emotion*. Edinburgh: Edinburgh University Press.

Arditti, Rita, Renate Duelli-Klein, and Shelley Minden, eds. 1984. *Test-Tube Women: What Future for Motherhood*. London: Pandora.

Armstrong, David. 1994. "Medical Surveillance of Normal Populations." In G. Lawrence, ed., *Technologies of Modern Medicine*. London: Science Museum.

———. 2001. "The Rise of Surveillance Medicine." In M. Purdy and D. Banks, eds., *The Sociology and Politics of Health: A Reader*. London: Routledge.

Asad, Talal. 1973. *Anthropology and the Colonial Encounter*. London: Ithaca Press.

Atkinson, Paul. 1995. *Medical Talk and Medical Work*. London: Sage.

Atkinson, Paul, Amanda Coffey, Sara Delamont, John Lofland, and Lyn Lofland, eds. 2001. *Handbook of Ethnography*. London: Sage.

Banchoff, Thomas. 2004. "Embryo Politics: Debating Life in a Global Era." Unpublished manuscript, cited with permission from the author.

Beck, Ulrich. 1992. *Risk Society: Towards a New Modernity*. Trans. from the German by Mark Ritter. London: Sage.

Becker, Gay. 2000. *The Elusive Embryo: How Men and Women Approach New Reproductive Technologies*. Berkeley: University of California Press.

Becker, Howard S. 1961. *Boys in White: Student Culture in Medical School*. Chicago: University of Chicago Press.

Bickerstaff, Helen, Frances Flinter, Cheng Toh Yeong, and Peter Braude. 2001. "Clinical Application of Preimplantation Genetic Diagnosis." *Human Fertility* 4:24–30.

Birke, Lynda, Susan Himmelweit, and Gail Vines. 1990. *Tomorrow's Child: Reproductive Technologies in the 90s*. London: Virago.

Bloor, Michael. 2001. "The Ethnography of Health and Medicine." In Paul Atkinson, Amanda Coffey, Sara Delamont, John Lofland, and Lyn Lofland, eds., *Handbook of Ethnography*. London: Sage, pp. 177–87.

Boseley, Sarah. 2001. "Fertility Authority Faces 'Designer Child' Decision." *Guardian*, October 2.

Bosk, Charles L. 1992. *All God's Mistakes: Genetic Counseling in a Pediatric Hospital*. Chicago and London: University of Chicago Press.

Bouquet, Mary. 1993. *Reclaiming English Kinship: Portuguese Refractions of British Kinship Theory*. Manchester: Manchester University Press.

Braude, Peter, Susan Pickering, Frances Flinter, and Caroline Mackie Ogilvie. 2002. "Preimplantation Genetic Diagnosis." *Nature Review Genetics* 3 (12): 941–53.

Brodwin, Paul E., ed. 1999. *Biotechnology and Culture: Bodies, Anxieties, Ethics*. Bloomington and Indianapolis: Indiana University Press.

Brown, Nik. 2003. "Hope against Hype: Accountability in Biopasts, Presents and Futures." *Science Studies* 16 (2): 3–21.

Brown, Nik, and Mike Michael. 2002. "From Authority to Authenticity: The Changing Governance of Biotechnology." *Health, Risk and Society* 4 (3): 259–72.

———. 2003. "A Sociology of Expectations: Retrospecting Prospects and Prospecting Retrospects." *Technology and Strategic Management* 15 (1): 3–18.

Brown, Nik, and Andrew Webster. 2004. *New Medical Technologies and Society: Reordering Life*. Cambridge: Polity.

Buster, J. E., and S. A. Carson. 1989. "Genetic Diagnosis of the Preimplantation Embryo." *American Journal of Medical Genetics* 34:211–16.

Caplan, Pat, ed. 2000. *Risk Revisited*. London: Pluto Press.

Carsten, Janet, ed. 2000. *Cultures of Relatedness: New Approaches to the Study of Kinship*. Cambridge: Cambridge University Press.

———. 2004. *After Kinship*. Cambridge: Cambridge University Press.

Challoner, Jack. 1999. *The Baby Makers: The History of Artificial Conception*. Basingstoke and London: Macmillan.

Clarke, A. J., and E. P. Parsons, eds. 1997. *Culture, Kinship and Genes: Towards Cross-Cultural Genetics*. Basingstoke: Macmillan.

Clifford, J., and G. E. Marcus, eds. 1986. *Writing Culture: The Poetics and Politics of Ethnography*. Cambridge: Cambridge University Press.

Condit, Celeste. 2004. "The Meanings and Effects of Discourse about Genetics: Methodological Variations in Studies of Discourse and Social Change." *Discourse and Society* 15 (4): 391–407.

Conrad, Peter, and Jonathan Gabe, eds. 1999. *Sociological Perspectives on the New Genetics*. Oxford: Blackwell.

Corea, Gena. 1985. *The Mother Machine: Reproductive Technologies from Artificial Insemination to Artificial Wombs*. New York: Harper & Row.

Corrigan, Oonagh, and Richard Tutton, eds. 2004. *Genetic Databases: Socioethical Issues in the Collection and Use of DNA*. London: Routledge.

Cram, David, and David de Kretser. 2002. "Genetic Diagnosis: The Future." In Christopher J. De Jonge and Christopher L. R. Barratt, eds., *Assisted Repro-*

ductive Technology: Accomplishments and New Horizons. Cambridge: Cambridge University Press, pp.186–205.

Crowe, Christine. 1985. "'Women Want It': In Vitro Fertilisation and Women's Motivations for Participation." *Women's Studies International Forum* 8:547–52.

———. 1990. "Whose Mind over Whose Matter? Women, In Vitro Fertilisation and the Development of Scientific Knowledge." In Maureen McNeil, Ian Varcoe, and Steven Yearley, eds., *The New Reproductive Technologies*. London: Macmillan, pp. 27–57.

Cussins, Charis. 1996. "Ontological Choreography: Agency through Objectification in Infertility Clinics." *Social Studies of Science* 26 (3): 575–610

———. 1998. "Producing Reproduction: Techniques of Normalization and Naturalization in Infertility Clinics." In Sarah Franklin and Helena Ragone, eds., *Reproducing Reproduction: Kinship, Power, and Technological Innovation*. Philadelphia: University of Pennsylvania Press, pp. 66–101.

De Jonge, Christopher J., and Christopher L. R. Barratt, eds. 2002. *Assisted Reproductive Technology: Accomplishments and New Horizons*. Cambridge: Cambridge University Press.

Duden, Barbara, and Silja Samerski. 2003. "The Release of Genetic Terms into Everyday Language: How the 'Pop Gene' Stimulates Risk Anxiety." Paper presented at "Genes, Gender and Generation," Lancaster University, 26 June.

Duster, Troy. 2003. *Backdoor to Eugenics*. Second Edition. New York: Routledge.

Edwards, Jeanette. 2000. *Born and Bred: Idioms of Kinship and New Reproductive Technologies in England*. Oxford: Oxford University Press.

Edwards, Jeanette, Sarah Franklin, Eric Hirsch, Francis Price, and Marilyn Strathern. 1993. *Technologies of Procreation: Kinship in the Age of Assisted Conception*. Manchester: Manchester University Press.

———. 1999. *Technologies of Procreation: Kinship in the Age of Assisted Conception*. Second Edition. London: Routledge.

Edwards, R. G. 1965. "Maturation In Vitro of Mouse, Sheep, Cow, Pig, Rhesus Monkey and Human Ovarian Oocytes." *Nature* 208:349–51.

———. 1987. "Diagnostic Methods for Human Gametes and Embryos." *Human Reproduction* 2:415–20.

———, ed. 1993. *Preconception and Preimplantation Diagnosis of Human Genetic Disease*. Cambridge: Cambridge University Press.

———. 2001. "The Bumpy Road to Human In Vitro Fertilization." *Nature Medicine* 7 (10): 1091–94.

———. 2002. "Ethics of Preimplantation Diagnosis: Recordings from the Fourth International Symposium on Preimplantation Genetics." *Reproductive BioMedicine Online* 6 (2): 170–80.

———. 2004. "Stem Cells Today: A. Origin and Potential of Embryo Stem Cells." *Reproductive BioMedicine Online* 8 (3): 275–306.

———. 2005a. "Changing World of IVF, Stem Cells, and PGD." Keynote address presented at the Sixth International Symposium on Preimplantation Genetics, London, 19–21 May. See *Reproductive BioMedicine Online* 10 (2):1 (abstract O-2).

235

———. 2005b. "Introduction: The Beginnings of In-Vitro Fertilization and Its Derivatives." In *Reproductive BioMedicine Online* publications, Modern Assisted Conception. Cambridge: Reproductive Healthcare, pp. 1–7.

Edwards, R. G., and R. L. Gardner. 1967. "Sexing of Live Rabbit Blastocysts." *Nature* 214:576–77.

———. 1968. "Choosing Sex before Birth." *New Scientist* 38:218–20.

Edwards, R. G., and P. Holland. 1988. "New Advances in Human Embryology: Implications of the Preimplantation Diagnosis of Genetic Disease." *Human Reproduction* 3:549–56.

Edwards, R. G., and J. D. Shulman. 1993. "History of and Opportunities for Preimplantation Diagnosis." In R. G. Edwards, ed., *Preconception and Preimplantation Diagnosis of Human Genetic Disease*. Cambridge: Cambridge University Press, pp. 3–42.

Edwards, Robert G., and Patrick Steptoe. 1980. *A Matter of Life: the Story of a Medical Breakthrough*. London: Hutchinson.

Emerson, Robert M., Rachel I. Fretz, and Linda L. Shaw. 2001. "Participant Observation and Fieldnotes." In Paul Atkinson, Amanda Coffey, Sara Delamont, John Lofland, and Lyn Lofland, eds., *Handbook of Ethnography*. London: Sage.

ESHRE PGD Consortium Steering Committee. 2002. "ESHRE Preimplantation Genetic Diagnosis Consortium Data Collection III (2001)." *Human Reproduction* 17 (1): 233–46.

ESHRE Preimplantation Genetic Diagnosis Consortium. 2000. "Data Collection II (May 2000)." *Human Reproduction* 15:2673–83

———. 2002. "2002 Data Collection III." *Human Reproduction* 17:233–46.

Ettore, Elizabeth. 2002. *Reproductive Genetics, Gender and the Body*. London: Routledge.

Faubion, James D. 2001. "Currents of Cultural Fieldwork." In Paul Atkinson, Amanda Coffey, Sara Delamont, John Lofland, and Lyn Lofland, eds., *Handbook of Ethnography*. London: Sage.

Finkler, Kaja. 2000. *Experiencing the New Genetics: Family and Kinship on the Medical Frontier*. Philadelphia: University of Pennsylvania Press.

Flower, Michael J., and Deborah Heath. 1993. "Micro-Anatomo Politics: Mapping the Human Genome Project." *Culture, Medicine and Psychiatry* 17:27–41.

Foucault, Michel. 1981. *The History of Sexuality, Volume 1: An Introduction*. Trans. Michael Hurley. Harmondsworth: Penguin [1978].

Frank, Arthur. 2003. "Connecting Body Parts: Technolux, Surgical Shapings and Bioethics." Paper presented at "Vital Politics," London School of Economics, September.

Franklin, Sarah. 1992. "Making Sense of Missed Conceptions: Anthropological Perspectives on Unexplained Infertility." In Meg Stacey, ed., *Changing Human Reproduction: Social Science Perspectives*. London: Sage, pp. 75–91.

———. 1995a. "Science as Culture, Cultures of Science." *Annual Reviews of Anthropology* 24:163–84.

———. 1995b. "Postmodern Procreation: A Cultural Account of Assisted Reproduction." In Faye Ginsburg and Rayna Rapp, eds. *Conceiving the New World Order: The Global Politics of Reproduction*. Berkeley: University of California Press, pp. 323–345.

———. 1997. *Embodied Progress: A Cultural Account of Reproduction.* London: Routledge, 1997.

———. 1999a. "Dead Embryos: Feminism in Suspension." In M. Michaels and L. Morgan, eds., *Fetal Subjects, Feminist Positions.* Philadelphia: University of Pennsylvania Press, pp. 61–82.

———. 1999b. "Making Representations: The Parliamentary Debate of the Human Fertilisation and Embryology Act." In J. Edwards, S. Franklin, E. Hirsch, F. Price, and M. Strathern, *Technologies of Procreation: Kinship in the Age of Assisted Conception.* Second Edition. London: Routledge, pp. 127–65.

———. 1999c. "Orphaned Embryos." In J. Edwards, S. Franklin, E. Hirsch, F. Price, and M. Strathern, *Technologies of Procreation: Kinship in the Age of Assisted Conception.* Second Edition. London: Routledge, pp. 166–70.

———. 2001. "Culturing Biology: Cell Lines for the Second Millennium." *Health* 5 (3): 335–54.

———. 2003a. "Definitions of Genetic Knowledge in the Context of Preimplantation Genetic Diagnosis: An Ethnographic Study." End of Award Report, ESRC Award Reference Number L21825036.

———. 2003b. "Ethical Biocapital: New Strategies of Stem cell Culture." In S. Franklin and M. Lock, eds., *Remaking Life and Death: Towards an Anthropology of Biomedicine.* Santa Fe, NM: School of American Research Press, pp. 97–129.

———. 2003c. "Rethinking Nature/Culture: Anthropology and the New Genetics." *Anthropological Theory* 3 (1): 65–85.

———. 2003d. "Drawing the Line at Not-Fully Human: What We Already Know from Embryos." *American Journal of Bioethics* 3:3:W25–W27.

———. 2005a. "Consent Session Data Collection Pilot Exercise Results." Findings presented at the Second Meeting of the UK Network of hES Cell Coordinators (hESCCO), Leeds, 28–29 June.

———. 2005b. "The Reproductive Revolution: How Far Have We Come?" Professorial Inaugural Lecture, London School of Economics, 24 November (http://www.lse.ac.uk/collections/LSEPublicLecturesAndEvents/events/2005/2 0050919t1600z001.htm).

———. 2007. *Dolly Mixtures: The Remaking of Genealogy.* Durham, NC: Duke University Press.

Franklin, Sarah, and Margaret Lock. 2003a. "Animation and Cessation: The Remaking of Life and Death." In S. Franklin and M. Lock, eds., *Remaking Life and Death: Toward an Anthropology of the Biosciences.* Santa Fe, NM: School of American Research Press.

———, eds. 2003b. *Remaking Life and Death: Toward an Anthropology of the Biosciences.* Santa Fe, NM: School of American Research Press.

Franklin, Sarah, Celia Lury, and Jackie Stacey. 2000. *Global Nature, Global Culture.* London: Sage.

Franklin, Sarah, and Susan McKinnon. 2001a. "Introduction". In Sarah Franklin and Susan McKinnon, eds., *Relative Values: Reconfiguring Kinship Studies.* Durham, NC: Duke University Press.

———, eds. 2001b. *Relative Values: Reconfiguring Kinship Studies.* Durham, NC: Duke University Press.

Franklin, Sarah, Maureen McNeil, and Celia Roberts. 2002. "The Genetic Tie: A Cultural Account." Grant application submitted to the Economic and Social Research Council.

———. n.d. "The Baby with the Bathwater: Feminism and the Politics of Reproductive Citizenship." Under review.

Franklin, Sarah, and Helena Ragone. 1998a. "Introduction." In Sarah Franklin and Helena Ragone, eds., *Reproducing Reproduction: Kinship, Power, and Technological Innovation*. Philadelphia: University of Pennsylvania Press, pp. 1–14.

———, eds. 1998b. *Reproducing Reproduction: Kinship, Power, and Technological Innovation*. Philadelphia: University of Pennsylvania Press.

Franklin, Sarah, Celia Roberts, Karen Throsby, Peter Braude, Jenny Shaw, Alison Lashwood, and Sue Pickering. 2005. "Factors Affecting PGD Patients' Consent to Donate Embryos to Stem Cell Research." Paper presented at the Sixth International Symposium on Preimplantation Genetics, London, 19–21 May. (See "Conference Programme and Abstracts," volume 10, supplement 2, of Reproductive BioMedicine Online, p. 31.)

Franklin, Sarah, and Marilyn Strathern. 1992. *Kinship and New Genetic Technologies: An Assessment of Existing Anthropological Knowledge*. Report compiled for the Medical Research Division of the European Commission Human Genome Analysis Programme (DG-XII) under Ethical, Social and Legal Aspects of Human Genome Analysis, No. PL9101041.

Franklin, Sarah, and Richard Tutton. 2001. "Revisiting Concepts of Gift in the New Genetics." Department of Sociology, Lancaster University, and the Wellcome Trust.

Fukuyama, Francis. 2002. *Our Posthuman Future: Consequences of the Biotechnology Revolution*. New York: Farrar, Straus and Giroux.

Gardner, R. L. 1968. "Control of the Sex Ratio at Full Term in the Rabbit by Transferring Sexed Blastocysts." *Nature* 218 (139): 346–49.

Giddens, Anthony. 1991. *Modernity and Self-Identity: Self and Society in the Late Modern Age*. Cambridge: Polity.

Gillespie, Rosemary. 2003. "Childfree and Feminine: Understanding the Gender Identity of Voluntarily Childless Women." *Gender and Society* 17 (1): 122–36.

Ginsburg, Faye D., and Rayna Rapp, eds. 1995. *Conceiving the New World Order: The Global Politics of Reproduction*. Berkeley and London: University of California Press.

Glasner, Peter, and Harry Rothman, eds. 1998. *Genetic Imaginations: Ethical, Legal and Social Issues in Human Genome Research*. Aldershot: Ashgate.

Goodman, Alan H., Deborah Heath, and M. Susan Lindee, eds. 2003. *Genetic Nature/Culture: Anthropology and Science beyond the Two Culture Divide*. Berkeley: University of California Press.

Gosden, Roger. 1999. *Designing Babies: the Brave New World of Reproductive Technology*. New York: W. H. Freeman.

Graham, Chris. 2000. "Mammalian Development in the UK (1950–1995)." *International Journal of Developmental Biology* 44:51–55.

Gratzer, Walter. 2000. "Afterword." In James D. Watson, *A Passion for DNA: Genes, Genomes and Society*. Oxford: Oxford University Press.

Gunning, Jennifer. 2000. *Assisted Conception: Research, Ethics and Law*. Aldershot: Ashgate, 2000.

Habermas, Jürgen. 2003. *The Future of Human Nature*. Cambridge: Polity.

Hage, Ghassan. 2002. *Against Paranoid Nationalism: Searching for Hope in a Shrinking Society*. Melbourne: Pluto Press.

Haimes, Erica. 2002. "What Can Sociology Contribute to the Study of Ethics? Theoretical, Empirical and Substantive Considerations." *Bioethics* 16 (2): 89–113.

Handyside, Alan H. 1998. "Clinical Evaluation of Preimplantation Genetic Diagnosis." *Prenatal Diagnosis* 18:1345–48.

Handyside, Alan H., and Joy D. A. Delhanty. 1997. "Preimplantation Genetic Diagnosis: Strategies and Surprises." *Trends in Genetics* 13:270–75.

Handyside, A., J. Pattinson, R. Penketh, J. Delhanty, R. Winston, and H. Leese. 1990. "Pregnancies from Biopsied Human Preimplantation Embryos Sexed by Y-specific DNA Amplification." *Nature* 344:768–70.

Handyside, A. H., J. K. Pattinson, R.J.A. Penketh, J.D.A. Delhanty, R.M.L. Winston, and E.G.D. Tiddenham. 1989. "Biopsy of Human Preimplantation Embryos and Sexing by DNA Amplification." *Lancet* 1:347–49.

Handyside, Alan H., Paul N. Scriven, and Caroline Mackie Ogilvie. 1998. "The Future of Preimplantation Genetic Diagnosis." *Human Reproduction* 13 (4): 249–55.

Haraway, Donna. 1997. *Modest_Witness@Second Millennium.FemaleMan_Meets_OncoMouse: Feminism and Technoscience*. New York: Routledge.

Harper, Joyce C., Joy D. A. Delhanty, and Alan Handyside, eds. 2001. *Preimplantation Genetic Diagnosis*. Chichester: John Wiley and Sons.

Harris, John. 1985. The Value of Life: *An Introduction to Medical Ethics*. London: Routledge and Kegan Paul.

Hartouni, Valerie. 1997. *Cultural Conceptions: On Reproductive Technologies and the Remaking of Life*. Minneapolis: University of Minnesota Press.

Hayden, Corinne P. 1995. "Gender, Genetics, and Generation: Reformulating Biology in Lesbian Kinship." *Cultural Anthropology* 10 (1): 41–63.

Heath, Deborah, and Paul Rabinow. 1993. *Bio-Politics: The Anthropology of the New Genetics and Immunology*. Special issue of *Culture, Medicine and Psychiatry* 17 (1).

Hedgecoe, Adam. 2004. *The Politics of Personalised Medicine: Pharmacogenetics in the Clinic*. Cambridge: Cambridge University Press.

Holding, C., and M. Monk. 1989. "Diagnosis of Beta-Thalassaemia by DNA Amplification in Single Blastomeres from Mouse Preimplantation Embryos." *Lancet* 2:532–35.

Holloway, Richard. 1999. *Godless Morality: Keeping Religion Out of Ethics*. Edinburgh: Canongate.

Homans, Hilary, ed. 1985. *The Sexual Politics of Reproduction*. Aldershot: Gower.

House of Lords Science and Technology Committee. 2000. *Science and Society*. London: HMSO.

239

Human Fertilisation and Embryology Authority. 2002a. "HFEA Confirms That HLA Tissue Typing May Only Take Place When Preimplantation Genetic Diagnosis Is Required to Avoid a Serious Genetic Disorder." Press release, 1 August, http://www.hfea.gov.uk/PressOffice/Archive/43573563.

———. 2002b. "HFEA Licence Committee Approves Two Applications for Research on Human Embryos to Produce Stem Cell Lines." Press release, 28 February, http://www.hfea.gov.uk/PressOffice/Archive/.

Human Fertilisation and Embryology Authority and Advisory Committee on Genetic Testing. 1999. "Consultation Document on Preimplantation Genetic Diagnosis." http://www.doh.gov.uk/genetics/pgdprinciples.htm, accessed 18/01/2001.

Huxley, Aldous. 1932. *Brave New World*. London: Chatto and Windus.

Inhorn, Marcia C. 1994. *Quest for Conception: Gender, Infertility and Egyptian Medical Traditions*. Philadelphia: University of Pennsylvania Press.

———. 2003. *Local Babies, Global Science: Gender, Religion, and In Vitro Fertilization in Egypt*. New York: Routledge.

Inhorn, Marcia C., and Frank Van Balen, eds. 2002. *Infertility around the Globe: New Thinking on Childlessness, Gender and Reproductive Technologies*. Berkeley and London: University of California Press.

International Working Group on Preimplantation Genetics. 2001. "Preimplantation Genetic Diagnosis—Experience of Three Thousand Clinical Cycles." *Reproductive BioMedicine Online* 3:49–53.

Jackson, Emily. 2001. *Regulating Reproduction: Law, Technology and Autonomy*. Oxford: Hart.

Jasanoff, Sheila. 2005. *Designs on Nature: Science and Democracy in Europe and the United States*. Princeton, NJ: Princeton University Press.

Kahn, Susan. 2000. *Reproducing Jews: A Cultural Account of Assisted Conception in Israel*. Durham, NC: Duke University Press, 2000.

Keller, Evelyn Fox. 2000. *The Century of the Gene: Metaphors of Twentieth Century Biology*. Cambridge, MA, and London: Harvard University Press.

Kerr, Anne. 2004. *Genetics and Society: A Sociology of Disease*. London: Routledge.

Kerr, Anne, and Sarah Cunningham-Burley. 2000. "On Ambivalence and Risk: Reflexive Modernity and the New Human Genetics." *Sociology* 34 (2): 283–304.

Kerr, Anne, Sarah Cunningham-Burley, and Amanda Amos. 1998a. "Drawing the Line: An Analysis of Lay People's Discussions about the New Genetics." *Public Understanding of Science* 7 (2): 113–33.

———. 1998b. "Eugenics and the New Genetics in Britain: Examining Contemporary Professionals' 'Accounts.' " *Science, Technology and Human Values* 23 (2): 175–98.

———. 1998c. "The New Human Genetics and Health: Mobilising Lay Expertise." *Public Understanding of Science* 7 (1): 41–60.

Kerr, Anne, and Sarah Franklin. Forthcoming. "Genetic Ambivalence: Expertise, Uncertainty and Communication in the Context of the New Genetics." In Andrew Webster, ed., *Innovative Medical Technologies*.

Kerr, Anne, and Tom Shakespeare. 2002. *Genetic Politics: From Eugenics to Genome*. Cheltenham: New Clarion.

Kimbrell, Andrew. 1993. *The Human Body Shop: The Engineering and Marketing of Life*. London: Harper Collins.

Klein, Renate, ed. 1989. *Infertility: Women Speak Out about Their Experiences of Reproductive Medicine*. London: Pandora.

Konrad, Monica. 1998. "Ova Donation and Symbols of Substance: Some Variations on the Theme of Sex, Gender and the Partible Body." *Journal of the Royal Anthropological Institute* (n.s.) 4:643–67.

———. 2002. "Pre-symptomatic Networks: Tracking Experts across Medical Science and the New Genetics." In C. Shore and S. Nugent, eds., *Elite Cultures: Anthropological Perspectives*. London: Routledge.

———. 2003a. "From Secrets of Life to the Life of Secrets: Tracing Genetic Knowledge as Genealogical Ethics in Biomedical Britain." *Journal of the Royal Anthropological Institute* 9 (2): 339–59.

———. 2003b. "Predictive Genetic Testing and the Makings of the Pre-symptomatic Person: Prognostic Moralities amongst Huntington's-affected Families." *Anthropology and Medicine* 10 (1): 23–49.

———. 2005a. *Nameless Relations: Anonymity, Melanesia, and Reproductive Gift Exchange between British Ova Donors and Recipients*. Oxford: Bergahn.

———. 2005b. *Narrating the New Predictive Genetics: Ethics, Ethnography and Science*. Cambridge: Cambridge University Press.

Kuliev, Anver, and Yury Verlinsky. 2002. "Current Features of Preimplantation Genetic Diagnosis." *Reproductive BioMedicine Online* 5 (3): 294–61.

Layne, Linda, ed. 1999. *Transformative Motherhood: On Giving and Getting in a Consumer Culture*. New York: New York University Press.

———. 2003. *Motherhood Lost: A Feminist Account of Pregnancy Loss in America*. New York: Routledge.

Letherby, Gayle. 1994. "Mother or Not, Other or What? Problems of Definition and Identity." *Women's Studies International Forum* 17 (5): 525–32.

———. 1999. "Other Than Mother and Mothers as Others: The Experience of Motherhood and Non-motherhood in Relation to 'Infertility' and 'Involuntary Childlessness.'" *Women's Studies International Forum* 22 (3): 359–72.

Lindenbaum, Shirley, and Margaret Lock, eds. 1993. *Knowledge, Power and Practice: The Anthropology of Medicine and Everyday Life*. Berkeley and London: University of California Press.

Lippman, A. 1992. "Led (Astray) by Genetic Maps: The Cartography of the Human Genome and Human Health Care." *Social Science and Medicine* 35 (12): 1469–76.

Lock, Margaret. 2001. *Twice Dead: Organ Transplants and the Reinvention of Death*. Berkeley: University of California Press.

———. 2005 "The Eclipse of the Gene and the Return of Divination." *Current Anthropology* 46:S47–S71.

Lock, Margaret, and Deborah R. Gordon, eds. 1988. *Biomedicine Examined*. Dordrecht and Boston: Kluwer Academic Publisher, 1988.

Lupton, Deborah. 1993. "Risk as Moral Danger: The Social and Political Functions of Risk Discourse in Public Health." *International Journal of Health Services* 23 (3): 425–35.

241

Marcus, George E. 1995. "Ethnography in/of the World System: The Emergence of Multisited Ethnography." *Annual Review of Anthropology* 24:95–117.

Marks, Jonathan. 1995. *Human Biodiversity: Genes, Race, and History.* New York: Aldine de Gruyter.

———. 2002. *What It Means to Be 98% Chimpanzee.* Berkeley: University of California Press.

Marteau, Theresa, and Martin Richards, eds. 1996. *The Troubled Helix: Social and Psychological Implications of the New Human Genetics.* Cambridge: Cambridge University Press.

McAllister, Fiona, and Linda Clarke. 1998. *Choosing Childlessness: Family and Parenthood, Policy and Practice.* London: Family Policy Studies Centre.

McKibben, Bill. 2003. *Enough: The Dangers of Being Superhuman.* London: Bloomsbury.

McLaren, Anne. 1985. "Prenatal Diagnosis before Implantation: Opportunities and Problems." *Prenatal Diagnosis* 5:85–90.

———. 1987. "Can We Diagnose Genetic Disease in Preembryos?" *New Scientist* 116 (10 Dec.), 42–47.

———. 2000. "Cloning: Pathways to a Pluripotent Future." *Science* 288 (5472): 1775–80.

McNeil, Maureen. 1987. *Under the Banner of Science: Erasmus Darwin and His Age.* Manchester: Manchester University Press.

McNeil, Maureen, Ian Varcoe, and Steven Yearley, eds. 1990. *The New Reproductive Technologies.* London: Macmillan.

Mitchell, Lisa M. 2001. *Baby's First Picture: Ultrasound and the Politics of Fetal Subjects.* Toronto: University of Toronto Press.

Modell, Judith S. 1994. *Kinship with Strangers: Adoption and Interpretations of Kinship in American Culture.* Berkeley: University of California Press.

Mol, Annemarie. 2003. *The Body Multiple: Ontology in Medical Practice.* Durham, NC, and London: Duke University Press.

Monk, M. 1988. "Pre-implantation Diagnosis." *BioEssays* 8:184–89.

———. 1990a. "Diagnosis of Disease in Preimplantation Embryos." In Nobel Prize Symposium, *The Aetiology of Human Disease at the DNA Level.* New York: Raven Press.

———. 1990b. "Preimplantation Diagnosis by Biochemical or DNA Microassay in a Single Cell." In R. G. Edwards, ed., *Establishing a Successful Human Pregnancy.* New York: Raven Press, pp. 183–97.

———. 1991. "Preimplantation Diagnosis." *Bibliography of Reproduction* 57 (2): A1–A8.

———. 2001. "Of Microbes, Mice and Man." *Mammalian Reproduction and Development* 45 (3): 231–45.

Monk, M., A. Handyside, K. Hardy, and D. Whittingham. 1987. "Preimplantation Diagnosis of Deficiency or Hypoxanthine Phosphoribosyl Transferase in a Mouse Model for Lesch-Nyhan Syndrome." *Lancet* 2:423–26.

Monk, M., and M. Harper. 1979. "Sequential X-chromosome Inactivation Linked to Cellular Differentiation in Early Mouse Development." *Nature* 281: 311–13.

Monk, M., and C. Holding. 1990. "Amplification of a B-haemoglobin Sequence in Individual Human Oocytes and Polar Bodies." *Lancet* 335:985–88.

Monk, M., and H. Kathuria. 1977. "Dosage Compensation of an X-linked Gene in Preimplantation Mouse Embryos." *Nature* 270:599–601.

Morgan, Derek, and Robert Lee. 1991. *Human Fertilisation and Embryology Act 1990: Abortion and Embryo Research, the New Law*. London: Blackstone Press, 1991.

———. 1997. "In the Name of the Father? *Ex parte Blood*: Dealing with Novelty and Anomaly." *Modern Law Review* 60 (69): 840–56.

Mulkay, Michael. 1993. "Rhetorics of Hope and Fear in the Great Embryo Debate." *Social Studies of Science* 23:721–42.

———. 1994. "Changing Minds about Embryo Research." *Public Understanding of Science* 3:195–213.

———. 1997. *The Embryo Research Debate: Science and the Politics of Reproduction*. Cambridge: Cambridge University Press.

Munne, S., C. Magli, J. Cohen, P. Morton, S. Sadowy, L. Gianaroli, M. Tucker, C. Marquez, D. Sable, A. P. Ferraretti, J. B. Massey, and R. Scott. 1999. "Positive Outcome after Preimplantation Diagnosis of Aneuploidy in Human Embryos." *Human Reproduction* 14 (9): 2191–99.

Murphy, Elizabeth, and Robert Dingwall. 2001. "The Ethics of Ethnography." In Paul Atkinson, Amanda Coffey, Sara Delamont, John Lofland, and Lyn Lofland, eds., *Handbook of Ethnography*. London: Sage.

Nash, Catherine. 2004. "Genetic Kinship." *Cultural Studies* 18 (1): 1–33.

Nelkin, Dorothy, and M. Susan Lindee. *The DNA Mystique: The Gene as a Cultural Icon*. New York: W. H. Freeman.

Nelkin, Dorothy, and Lawrence Tancredi. 1989. *Dangerous Diagnostics: the Social Power of Biological Information*. New York: Basic Books.

Novas, Carlos, and Nikolas Rose. 2000. "Genetic Risk and the Birth of the Somatic Individual." *Economy and Society* 29 (4): 484–513.

Oakley, Ann. 1980. *Women Confined: Towards a Sociology of Childbirth*. Oxford: Martin Robertson.

———. 1984. *The Captured Womb: A History of the Medical Care of Pregnant Women*. Oxford: Blackwell.

O'Neill, Onora. 2002a. *Autonomy and Trust in Bioethics*. Cambridge: Cambridge University Press.

———. 2002b. *A Question of Trust: The BBC Reith Lectures 2002*. Cambridge: Cambridge University Press.

Penketh, R. 1993. "The Scope of Preimplantation Genetic Diagnosis." In Robert G. Edwards, ed., *Preconception and Preimplantation Diagnosis of Human Genetic Disease*. Cambridge: Cambridge University Press, pp. 81–97.

Penketh, Richard, and Anne McLaren. 1987. "Prospects for Prenatal Diagnosis during Preimplantation Human Development." *Baillière's Clinical Obstetrics and Gynaecology* 1 (3): 747–64.

Pergament, Eugene. 2001. "Foreword." In Joyce C. Harper, Joy D. A. Delhanty, and Alan Handyside, eds., *Preimplantation Genetic Diagnosis*. Chichester: John Wiley and Sons.

243

Petersen, Alan, and Robin Bunton. 2002. *The New Genetics and the Public's Health*. London: Routledge.

Petryna, Adriana. 2002. *Life Exposed: Biological Citizens after Chernobyl*. Princeton, NJ: Princeton University Press.

Pfeffer, Naomi, and Anne Woollett. 1983. *The Experience of Infertility*. London: Virago.

Pickering, Susan, Peter Braude, Minal Patel, Chris J. Burns, Jane Trussler, Virginia Bolton, and Stephen Minger. 2003. "Preimplantation Genetic Diagnosis as a Novel Source of Embryos for Stem Cell Research." *Reproductive Bio-Medicine Online* 7 (3): 353–64.

Pickering, Susan J., Stephen L. Minger, Minal Patel, Hannah Taylor, Cheryl Black, Chris J. Burns, Antigoni Ekonomou, and Peter R. Braude. 2005. "Generation of a Human Embryonic Cell Line Encoding the Cystic Fibrosis Mutation ^F508, Using Preimplantion Genetic Diagnosis." *Reproductive BioMedicine Online* 10 (3): 390–97.

Pickering, Susan, Nikolaos Polidoropoulos, Jenny Caller, Paul Scriven, Caroline Mackie Ogilvie, Peter Braude, and the PGD Study Group. 2003. "Strategies and Outcomes of the First 100 Cycles of Preimplantation Genetic Diagnosis at the Guy's and St Thomas' Center." *Fertility and Sterility* 79 (1): 81–90.

Pilnick, Alison. 2002. *Genetics and Society: An Introduction*. Buckingham: Open University Press.

Pinker, Stephen. 2003. "The Designer Baby Myth." *Guardian*, Thursday 5 June.

Potamainou, Anna. 1997. *Hope: A Shield in the Economy of Borderline States*. London: Routledge.

Power, Michael. 1999. *The Audit Society: Rituals of Verification*. Oxford: Oxford University Press.

———. 2004. *The Risk Management of Everything: Rethinking the Politics of Uncertainty*. London: Demos.

Rabinow, Paul. 1992. "Artificiality and Enlightenment: From Sociobiology to Biosociality." In Jonathan Crary and Sanford Kwinter, eds., *Incorporations*. New York: Zone Books, pp. 234–52.

———. 1996. *Making PCR: A Story of Biotechnology*. Chicago: University of Chicago Press.

———. 1997. *Essays on the Anthropology of Reason*. Princeton, NJ: Princeton University Press.

———. 1999. *French DNA: Trouble in Purgatory*. Chicago: Chicago University Press.

———. 2003. *Anthropos Today: Reflections on Modern Equipment*. Princeton, NJ: Princeton University Press.

Ragone, Helena. 1994. *Surrogate Motherhood: Conception in the Heart*. Boulder and Oxford: Westview Press.

Rapp, Rayna. 1999. *Testing Women, Testing the Fetus: The Social Impact of Amniocentesis in America*. New York and London: Routledge.

———. 2003. "Cell Life and Death, Child Life and Death: Genomic Horizons, Genetic Disease, Family Stories' Culture." In S. Franklin and M. Lock, eds., *Remaking Life and Death: Toward an Anthropology of Biomedicine*. Santa Fe, NM: School of American Research Press, pp. 129–64.

Rapp, Rayna, and Faye Ginsburg. 2002. "Enabling Disability: Rewriting Kinship, Reimagining Citizenship." *Public Culture* 13 (3): 553–66.

Ray, P. F., R.M.L. Winston, and A. H. Handyside. "XIST Expression from the Maternal X Chromosome in Human Male Preimplantation Embryos at the Blastocyst Stage." *Human Molecular Genetics* 99:1323–27.

Redfield, Peter. 2000. *Space in the Tropics: From Convicts to Rockets in French Guiana*. Berkeley: University of California Press.

Rose, Nikolas. 2001. "The Politics of Life Itself." *Theory, Culture and Society* 18:1–30.

———. 2006. *The Politics of Life Itself*. Princeton, NJ: Princeton University Press.

Rose, Nikolas, and Carlos Novas. 2005. "Biological Citizenship." In Aihwa Ong and Stephen J. Collier, eds., *Global Assemblages: Technology, Politics, and Ethics as Anthropological Problems*. Malden: Blackwell, pp. 439–63.

Rothman, Barbara Katz. 1986. *The Tentative Pregnancy: Prenatal Diagnosis and the Future of Motherhood*. New York: Viking.

———. 1994. *The Tentative Pregnancy: Amniocentesis and the Sexual Politics of Motherhood*. London: Pandora.

Sandelowski, Margarete. 1993. *With Child in Mind: Studies of the Personal Encounter with Infertility*. Philadelphia: University of Pennsylvania Press, 1993.

Sayers, Sean. 2001. "Unbounded Justice." Review of Onora O'Neill's *Bounds of Justice*, www.theglobalsite.ac.uk, accessed 11/12/2003.

Schaffer, Steven, and Simon Shapin. 1989. *Leviathan and the Air-Pump: Hobbes, Boyle and the Experimental Life*. Princeton, NJ: Princeton University Press.

Scriven, Paul, Alan H. Handyside, and Caroline Mackie Ogilvie. 1998. "Chromosome Translocations: Segregation Modes and Strategies for Preimplantation Genetic Diagnosis." *Prenatal Diagnosis* 18:1437–49.

Shakespeare, T. 2003. "Rights, Risks and Responsibilities: New Genetics and Disabled People." In S. J. Williams, L. Birke, and G. A. Bendelow, eds., *Debating Biology: Sociological Reflections on Health, Medicine and Society*. London: Routledge, pp. 198–209.

Shakespeare, Tom, and Anne Kerr. 1999. *Genetic Politics: From Eugenics to Genome*. Cheltenham: New Clarion.

Shelley, Mary. 1994. *Frankenstein; or, The Modern Prometheus*. London: Penguin.

Silver, Lee M. 1999. *Remaking Eden: Cloning, Genetic Engineering and the Future of Humankind?* New York: Phoenix.

Silverman, Sydel. 2003. "Foreword." In Alan H. Goodman, Deborah Heath, and M. Susan Lindee, eds., *Genetic Nature/Culture: Anthropology and Science beyond the Two-Culture Divide*. Berkeley: University of California Press, pp. 1x–xv.

Simpson, Bob. 2000. "Imagined Genetic Communities: Ethnicity and Essentialism in the Twenty-First Century." *Anthropology Today* 16 (3): 3–6.

———. 2001. "Making 'Bad' Deaths 'Good': The Kinship Consequences of Posthumous Conception." *Journal of the Royal Anthropological Institute* 7: 1–18 (intro).

Sloterdijk, Peter. 2000. *Règles pour le parc humain: Une lettre en réponse à la Lettre sure l'humanisme de Heidegger*. Trans. Olivier Mannoni. Paris: Editions Mille et Une Nuits.

Spallone, Patricia. 1996. "The Salutary Tale of the Pre-Embryo." In Nina Lykke and Rosi Braidotti, eds., *Between Monsters Goddesses and Cyborgs: Feminist Confrontations with Science, Medicine and Cyberspace*. London: Palgrave, Macmillan.

Spallone, Patricia, and Deborah Lynn Steinberg, eds. 1987. *Made to Order: The Myth of Reproductive and Genetic Progress*. Oxford: Pergamon.

Squier, Susan. 1994. *Babies in Bottles: Twentieth Century Visions of Reproductive Technology*. New Brunswick, NJ: Rutgers University Press.

Stacey, Jackie. 2005. "Masculinity, Masquerade, and Genetic Impersonation: Gattaca's Queer Visions." *Signs* 30 (3): 1851–77.

Stacey, Judith. 1996. *In The Name of the Family: Rethinking Family Values in the Postmodern Age*. Boston: Beacon Press.

———. 1998. *Brave New Families: Stories of Domestic Upheaval in Late Twentieth Century America*. Berkeley: University of California Press.

Stacey, Meg, ed. 1992. *Changing Human Reproduction: Social Science Perspectives*. London, Newbury Park and New Delhi: Sage.

Stanworth, Michelle, ed. 1987. *Reproductive Technologies: Gender, Motherhood and Medicine*. Cambridge: Polity.

Steinberg, Deborah Lynn. 1997. *Bodies in Glass: Genetics, Eugenics, Embryo Ethics*. Manchester: Manchester University Press.

Stock, Gregory. 2002. *Redesigning Humans: Our Inevitable Genetic Future*. Boston and New York: Houghton Mifflin.

Stocking, George, ed. 1983. *Observers Observed: Essays on Ethnographic Fieldwork*. Madison: University of Wisconsin Press.

Strathern, Marilyn. 1992a. *After Nature: Kinship in the Late Twentieth Century*. Cambridge: Cambridge University Press.

———. 1992b. *Reproducing the Future: Anthropology, Kinship and the New Reproductive Technologies*. Manchester: Manchester University Press.

———. 1999. "Regulation, Substitution and Possibility." In Jeanette Edwards, Sarah Franklin, Eric Hirsch, Frances Price, and Marilyn Strathern, *Technologies of Procreation: Kinship in the Age of Assisted Conception*, second edition. London: Routledge, pp. 171–202.

———, ed. 2000. *Audit Cultures: Anthropological Studies in Accountability, Ethics and the Academy*. London: Routledge.

———. 2005. *Kinship, Law and the Unexpected: Relatives Are Always a Surprise*. Cambridge: Cambridge University Press.

Thompson, Charis. 2001. "Strategic Naturalizing: Kinship in an Infertility Clinic." In Sarah Franklin and Susan McKinnon, eds., *Relative Values: Reconfiguring Kinship Studies*. Durham, NC: Duke University Press, pp. 175–202.

———. 2005. *Making Parents: The Ontological Choreography: Reproductive Technologies*. Cambridge: MIT Press.

Throsby, Karen. 2004. *When IVF Fails: Feminism, Infertility and the Negotiation of Normality*. London: Palgrave.

Turner, Victor. 1969. *The Ritual Process: Structure and Anti-structure*. London: Routledge and Kegan Paul.

Turney, Jon. 1998. *Frankenstein's Footsteps: Science, Genetics and Popular Culture*. New Haven, CT, and London: Yale University Press.

Tutton, Richard. 2002. "Gift Relationships in Genetics Research." *Science as Culture* 11 (4): 523–42.

Tutton, Richard, and Oonagh Corrigan, eds. 2004. *Genetic Databases: Socio-Ethical Issues in the Collection and Use of DNA*. London: Routledge.

Tyler, Imogen. 2001. "Skin-Tight: Celebrity, Pregnancy and Subjectivity." In S. Ahmed and J. Stacey, eds. *Thinking through the Skin*. London and New York: Routledge.

Van Dijck, José. 1998. *Imagenation: Popular Images of Genetics*. Basingstoke: Macmillan.

Verlinsky, Yuri, and Anver Kuliev. 1998. "Preimplantation Genetics." *Journal of Assisted Reproduction and Genetics* 15 (5): 215–18.

———. 2000. *An Atlas of Preimplantation Genetic Diagnosis*. New York and London: Parthenon.

Warnock, Mary. 1985. *A Question of Life: The Warnock Report on Human Fertilisation and Embryology*. Oxford: Blackwell.

———. 2000. *A Memoir*. London: Duckworth.

———. 2002. *Making Babies: Is There a Right to Have Children?* Oxford: Oxford University Press.

———. 2003. *Nature and Morality: Recollections of a Philosopher in Public Life*. London: Continuum.

Watson, James. 1968. *The Double Helix: A Personal Account of the Discovery of the Structure of DNA*. New York: Mentor.

———. 2000. *A Passion for DNA: Genes, Genomes and Society*. Oxford: Oxford University Press.

———. 2005. "Human Genetic Variation." Keynote address presented at the Sixth International Symposium on Preimplantation Genetics, London, 19–21 May. See *Reproductive BioMedicine Online* 10 (2): 1 (abstract O-1).

Weiner, Annette. 1995. "Culture and Our Discontents." *American Anthropologist* 97 (1): 14–40.

Whittingham, D. G., and R. Penketh. 1987. "Prenatal Diagnosis in the Human Preimplantation Period." *Human Reproduction* 2:267–70.

Wild, Leah. 2000. "Would I Be Eliminated?" *Guardian*, 29 June, p. 16.

Willadsen, S. 1979. "A Method for the Culture of Micromanipulated Sheep Embryos and Its Use to Provide Monozygotic Twins." *Nature* 277:289–300.

———. 1980. "The Viability of Early Cleavage Stages Containing Half the Number of Blastomeres in the Sheep." *Journal of Reproduction and Fertility* 59:357–62.

———. 1986. "Nuclear Transplantation in Sheep Embryos." *Nature* 320:63–65.

Wilmut, Ian, Keith Campbell, and Colin Tudge. 2000. *The Second Creation: The Age of Biological Control by the Scientists Who Cloned Dolly*. London: Headline.

Winston, Robert. 1999. *The IVF Revolution: The Definitive Guide to Assisted Reproductive Techniques*. London: Vermillion.

Zeiler, Kristin. 2004. "Reproductive Autonomous Choice—a Cherished Illusion?

REFERENCES

Reproductive Autonomy Examined in the Context of Preimplantation Genetic Diagnosis." *Medicine, Healthcare and Philosophy* 7:175–83.

———. 2005. "Chosen Children: An Empirical Study and a Philosophical Analysis of Moral Aspects of Pre-implantation Genetic Diagnosis and Germ-Line Gene Therapy." Doctoral dissertation submitted to the Department of Health and Society, Linkopings Universitet, Linkoping, Sweden.

248

Index

FORMATION *Series*

Everything Was Forever, Until It Was No More:
The Last Soviet Generation
BY ALEXEI YURCHAK

Wild Profusion: Biodiversity Conservation in an Indonesian Archipelago
BY CELIA LOWE

Born and Made: An Ethnography of Preimplantation Genetic Diagnosis
BY SARAH FRANKLIN AND CELIA ROBERTS